Freshwater fishes in Britain

CEH
Centre for
Ecology & Hydrology
NATURAL ENVIRONMENT RESEARCH COUNCIL

ENVIRONMENT
AGENCY

JOINT
NATURE
CONSERVATION
COMMITTEE

Freshwater fishes in Britain

the species and their distribution

Compiled and edited by

Cynthia E. Davies, Jonathan Shelley,
Paul T. Harding,
Ian F. G. McLean, Ross Gardiner
and Graeme Peirson

With assistance from

Vicky Quill and the authors of
individual species accounts

A joint project of the

Centre for Ecology and Hydrology,
Environment Agency
and
Joint Nature Conservation Committee

2004

Harley Books (B. H. & A. Harley Ltd)
Martins, Great Horkesley,
Colchester, Essex CO6 4AH, England

Website: www.harleybooks.com

Text set in Sabon and designed by James Shurmer

Text printed by St Edmundsbury Press Ltd, Bury St Edmunds, Suffolk

Colour reproduced and printed by Hilo Colour Printers Ltd, Colchester, Essex

The vignettes on pages 2, 8, 18, 147 and 162 are taken from W. Yarrell's *A History of British Fishes*, Edn 3, 1859.

Cover photo of Brown Trout in the River Test near Whitchurch, Hampshire © Dorian Moss

British Freshwater Fishes – the species and their distribution

published by Harley Books, 2004

British Library Cataloguing-in-Publication Data applied for

ISBN 0 946589 76 3

Contents

Contents *continued*

Foreword

For many, fishes are mysterious, almost alien creatures, living as they do in an aquatic environment wholly distinct from that of the majority of living organisms, including ourselves. Perhaps, subconsciously, that is part of their undeniable attraction for our species. The human race is innately inquisitive, so it would not be surprising that fishes should have attracted intense interest over many thousands of years.

The nature of that attraction is various. For most, their elegant, seemingly effortless motility, as well as the mystery of their lives in the unknown depths, excites both admiration and interest. It is not surprising, therefore, that there is an immense corpus of detailed folklore and knowledge of the habits and behaviour of fishes for every local pond, lake, stream or river that they inhabit. Of course, their edible qualities must have played some part also in their attraction, especially for our early ancestors, whose need to fish effectively led to increasing knowledge. Fishes were amongst the earliest migrants into Britain throughout the Quaternary upheavals. Thereafter, their distribution was further complicated by their importation and transfer to new sites as a result of human activities – a phenomenon that persists to this day.

Anyone seeking to learn more about fishes is faced with a vast body of writings, the majority concerned either with some technical aspect of fishing, now one of the most popular sports in Britain, or the extraordinary emotions and diversity of reflections that fishing induces among anglers! Within that voluminous literature, much of it repetitious, is an enormous body of information about fishes and their ways. Nevertheless, much is still unknown. In particular, it has proved difficult to obtain a clear picture of the general distribution of British fishes because most of the information, often immensely detailed, is held locally.

This atlas breaks new ground, or perhaps better, penetrates new waters! It provides the most comprehensive and authoritative account of the general distribution of the freshwater fishes of Britain. Although the distribution maps are the most original elements, the work provides much more. Each species is mapped on a 10km grid and is accompanied by a brief description of its essential characteristics, a short account of its biology, behaviour, habitat features, British and world distribution, biogeographic status, taxonomic status and possible hybridization, its relationship with people and, finally, a selected bibliography of further reading. The individual accounts are preceded by a general account of fish distribution in the British Isles and how it has come about, together with an account of the methodology adopted in the preparation of the distribution maps. All this represents the admirably edited contributions of many observers combined in a well-defined and reliable manner, such as one has come to expect from the Biological Records Centre at Monks Wood. The work concludes with a brief but highly apposite account of the issues involved in the conservation of British freshwater fishes, together with a full bibliography, a list of relevant web sites and details of legislation affecting British freshwater fishes.

Relatively ignorant of British fishes myself, I found this book of absorbing interest not only for the information it contained but also for the numerous opportunities it provides to follow up different topics or information on particular species. It is a splendid example of how basic distributional data allied with other information can illuminate our understanding of a group of organisms and their way of life. I congratulate the compilers of this Atlas for this splendid account which will be an indispensable baseline for current and future generations of fish biologists, amateurs and professionals alike, as well as those who are simply fascinated by fishes.

I commend it to all naturalists, fishermen, conservationists and those in government or its agencies who can affect the life of our freshwater fishes, as a unique, reliable source of clear and comprehensive information that is pleasing both to the mind and to the eyes. I believe it will lead to renewed enquiries into British freshwater fishes and, hopefully, to the more effective conservation of this 'alien race' in our midst.

December 2003

Sir John Burnett
Chairman
National Biodiversity Network Trust

Preface

Of making many books there is no end ... Ecclesiastes 12:12

Do we really need another book about freshwater fishes in Britain? In most bookshops and public libraries there is a section on fishes and fishing, usually with much of it devoted to freshwater species. Bookshops dealing with publishers' remainders or second-hand books are often particularly well stocked with books about fishes, written mainly for anglers. This suggests that there may be more books than potential readers! There are books on how to fish, where to fish, how enjoyable fishing is, fishing tales (tall and otherwise) and how to identify fishes. But none of these books deals with the detailed occurrence of our freshwater fishes from the perspective of the general naturalist.

Malcolm Greenhalgh, a self-confessed enthusiast for fishes and fishing, states that '*fish are "unattractive" animals*'. For him to make such a statement in his book *Freshwater fish* (1999) underlines an issue that has troubled us in writing our book. Except for anglers, those concerned with managing rivers and other waterbodies, and a few research scientists, British freshwater fishes now scarcely exist for the general public. Even children catching 'tiddlers'or sticklebacks in the local pond or canal, to take home in a jam jar, seem to be a thing of the past. Paradoxically, exotic fishes in expensively equipped tanks are popular pets of considerable commercial importance in retailing and publishing. The nearest most of the rest of us get to a freshwater fish now is via the supermarket fish counter, where farmed salmon and rainbow trout are plentiful and cheap. It is difficult to see living fishes in their natural habitat. But when you do it is often memorable: the synchronized swimming of a shoal of fry in the shallows, the bronze flanks of a bream breaking the surface of a calm river on a summer's morning, or the heroic leap of a sea trout running upstream to spawn.

By describing the distribution, roles and importance of freshwater fishes in Britain, we intend this book to make them more accessible and readily understood by the general naturalist and thereby to make them less 'unattractive'.

December 2003

Cynthia Davies, *Centre for Ecology and Hydrology*
Jonathan Shelley, *Environment Agency*
Paul Harding, *Biological Records Centre*
Ian McLean, *Joint Nature Conservation Committee*
Ross Gardiner, *Fisheries Research Services*
Graeme Peirson, *Environment Agency*

Acknowledgements

We are grateful to the Centre for Ecology and Hydrology (CEH), the Joint Nature Conservation Committee and the Environment Agency for funding the Database and Atlas of Freshwater Fishes (DAFF) project on which this book has been based.

Plans for such a project were first discussed by Paul Harding and Dr Ian Winfield over breakfast at a CEH conference on 15 August 1995, thanks to a timely introduction by Professor Alan Pickering (then Director at CEH Windermere). Ian helped prepare the feasibility study in 1996 and has generously provided support and advice many times throughout the project.

The maps have been derived from the DAFF database and the authors are most grateful to all the individuals and organizations, too numerous to list here, that have made datasets available to the project. All the data owners are acknowledged in the database. Particular thanks go to the Scottish Fisheries Trusts and Fishery Boards for access to fish distribution data for Scotland; to the many Environment Agency staff who gave both their time and effort to identify and supply data on behalf of the Agency; to staff of the Scottish Fisheries Co-ordination Centre; to Willie Duncan and Piers Langhelt for access to data; and also to Peter Maitland for access to data, including historical records, used in his 1972 *Key to British freshwater fishes*.

Staff at the Biological Records Centre, CEH Monks Wood, have assisted the DAFF database manager, Cynthia Davies, in compiling the database on which this book is based: thanks are due to Jon Cooper for developing the technique for extracting records from published maps, and for the reformatting, validation and input of data; to Bill Meek and Richard Bullen for the reformatting and validation of electronic data sources; to Val Burton for her rapid and accurate computerization of written records; and to Henry Arnold for his invaluable assistance with the production of the maps. The distribution maps published here were prepared using the DMAP software written by Dr Alan Morton.

Special thanks are due to the authors of the species accounts and the other chapters, whose contributions place the distribution maps in context. The species account authors are David Aldridge, Miran Aprahamian, Mike Atkinson, Philip Bolton, Robin Burrough, Nick Bromidge, Matt Carter, Sarah Chare, Alan Churchward, Steve Colclough, Ian Cook, Richard Cove, Ian Davidson, Keith Easton, Paul Frear, Roger Handford, Paul Harding, Phil Hickley, Andrew Hindes, David Hopkins, Richard Horsfield, Jim Lyons, Robin Musk, Sarah Peaty, Graeme Peirson, Adrian Pinder, Jonathan Shelley, James Vickers, Andy Walker, Ian Welby, Ian Winfield and Willie Yeomans. The remaining chapters and appendices were written by Cynthia Davies, Paul Harding, Ian McLean and Jonathan Shelley all with assistance and additional material from the editorial team.

Staff from several organizations have commented on text or supplied additional material, in particular Ian Winfield (CEH, Lancaster), Phil Hickley (Environment Agency), Dick Shelton (formerly of the Fisheries Research Services), Tristan Hatton-Ellis (Countryside Council for Wales), Colin Bean (Scottish Natural Heritage) and Henry Arnold (CEH, Monks Wood).

Scottish Natural Heritage supported the collation of data in Scotland; the Natural History Museum Library Service and Cambridge University Library provided access to many of the publications listed in Appendix 1 and the Bibliography; the Linnean Society of London gave permission for illustrations from their copy of Francis Day's *Fishes of Great Britain and Ireland* to be scanned for use in this book; Michael J. Roberts illustrated the 13 species in Chapter 4 not described and figured in Day, skilfully emulating his style. We are grateful to Sankurie Pye and Geoff Swinney of the National Museums of Scotland for locating suitable source material for Dr Roberts.

Finally, we must express our gratitude to Basil and Annette Harley for their editorial comments and for seeing the book through to publication with their characteristic attention to detail, and to James Shurmer for his excellent work on the design of the book.

Picture Credits

The photographer of Plates 4, 8 and 10 was D. W. Hay; of Plate 6, T. G. McInnes; of Plate 7, I. S. McLaren; of Plate 9, P. Rowarth; and of Plates 11, 12, 14 and 16, P. Wakely.

1 Introduction

Like most authors, we make the claim that our book is particularly distinctive. It differs from other books on British freshwater fishes in that it is based on wholly new information obtained from the Database and Atlas of Freshwater Fishes (DAFF), a project initiated in 1996 which ran until 2002. This introductory chapter aims to set freshwater fishes in a wider context. The factors affecting the distribution of individual species are analysed and discussed in some detail in Chapter 2, which is then followed, in Chapter 3, by a description and explanation of the DAFF project. The main part of the book comprises Chapter 4, in which all species of freshwater fishes occurring in Britain are described, with up-to-date accounts and distribution maps for each. Why we need to be concerned about these species and their conservation is discussed in Chapter 5.

The importance of freshwater fishes

Heritage and biodiversity

Freshwater fishes are part of our natural heritage and, with a mixture of indigenous and introduced species, they form part of the biodiversity of Britain. However, the concepts of heritage and biodiversity are very recent when compared with more than 10,000 years of contact between fishes and humans. This contact has been mainly through exploitation by people, but at the same time they have played an important part in the dispersal of many of our fish species. Freshwater fishes are just as important components of our natural heritage and biodiversity as wild flowers, birds or butterflies.

Environmental quality

Awareness of changes in the quality and 'health' of our environment really arose in the nineteenth century when rapid increases in industrialization, urbanization and population were recognized as not without adverse effects on the quality of life. For example, the easiest way to get rid of liquid waste from centres of population and from industrial processes was to channel it into the nearest watercourse. The pollution by foundries, mills and factories and their associated human populations of important rivers such as the Trent, Don, Mersey, Calder, Aire and Clyde is well documented. This pollution inevitably led to a loss of natural fish populations. Disposal of human waste into open sewers has taken place since urbanization began. With the introduction of the flushing water closet in the mid-1800s, even more human waste was washed into watercourses, leading to rapid reduction in the quality of rivers running through or downstream from larger towns and cities. The progressive loss of commercial fisheries along the tidal Thames, such as for eels, salmon, twaite shad and river lampreys, was due to pollution. It went almost unnoticed by politicians, until the Great Stink of 1858 penetrated the windows of Parliament. Reform of sewage legislation eventually followed in the Rivers Pollution Prevention Act, 1876. The Salmon and Freshwater Fisheries Act, 1923, recognized for the first time the importance of fishes and angling in the context of the need for cleaner rivers. During at least the last 80 years the angling lobby has done much to heighten awareness of the problems of water quality and the condition of our rivers, lakes and canals.

Recreation, sport and commercial interests

Fishing as a recreational activity has been known since at least the Middle Ages and probably has no more eloquent advocates than Izaak Walton in *The Compleat Angler*, and many of the writers included in Jeremy Paxman's anthology *Fish, Fishing and the Meaning of Life*. Fishing is now the most popular field sport in Britain (Plates 1 and 2), with thousands of angling clubs and more than three million people participating. Some clubs have only a handful of members, but others, such as the Birmingham Anglers Association, have over 15,000.

The Environment Agency has more than 450 employees responsible for maintaining, improving and developing fisheries in England and Wales (Plate 3). Recreational anglers in England and Wales alone spend in the region of £2.5 billion annually on their sport, and up to 50,000 jobs may depend wholly or partly on angling. The market value of fishing rights in England and Wales is estimated to be around £3 billion. Similar information is not yet available for Scotland, but the Scottish Executive has commissioned a study into the economic value of all forms of angling in Scotland which is currently in progress. However, the economic value in Scotland of the angling fishery for salmon alone was estimated in 1995 to be between £270 million and £430 million.

Most fishing, even for trout and salmon, is recreational (Plate 4), although individual fishes of some species are taken for food. The only commercially viable fisheries, based on wild stocks, are those for the netting or trapping of eels, salmon, sea trout and smelt. The economic value of these fisheries is limited nowadays, but can be locally important at times. The commercial elver fishery in the Severn Estuary exported elvers valued at about £1.5 million per annum in 1997/98, but this has declined rapidly in subsequent years due to world market forces. In Scotland, wild salmon and sea trout fisheries contribute significantly to the local economy, especially in regions such as the Tweed, Tay, Dee and Spey where recreational rod and line catches predominate over the commercial net-based fisheries found in coastal areas and estuaries (Plate 5). Results of a questionnaire survey in Scotland in 2002 showed that over 57,000 wild salmon (190.1 metric tonnes) and over 32,000 sea trout (36.5 metric tonnes) were caught by all methods and retained. Although commercial net and trap fisheries are in decline in Scotland, in 2002 they accounted for 41.1 per cent of the individual wild salmon and grilse caught and retained and 27.3 per cent of the sea trout. A small commercial fishery for smelt (usually known in Scotland as sparling), with about five boats operating part-time, exists in the Tay and Forth estuaries.

Fishes and other wildlife

Fishes as predators

The majority of British freshwater fishes feed mainly on invertebrates, but many species also eat a variety of vertebrate material, if only in the form of the eggs and fry of fishes, as well as algae and macrophytes. The selective feeding of some species is quite marked, but the season, location and age or size of the fish are important factors in determining the composition of the diet of individual species. Species such as trout and pike, when feeding selectively on either invertebrates or small fishes, exhibit different rates of growth depending on their prey. A comparison of the prey items of eels showed that in the River Cam in Cambridgeshire 22 distinct categories, from algae to voles, have been recorded, but in the Dubh Lochan in Stirlingshire there were only five – various invertebrates and newts. However, the methods used to describe diets in feeding studies vary greatly and comparisons of total biomass or nutritional value are rarely made. Freshwater fishes occupy an important position in freshwater food chains. Detailed studies have been made of fishes, invertebrates, macrophytes and algae in relation to nutrient cycles and energy flow in natural communities, for example at Windermere in Cumbria.

Fishes as prey

Freshwater fishes themselves form the prey of other fishes and of many other vertebrates. Almost all British species will take other fishes as prey, even if only as eggs or larvae. A few, for example pike, zander and perch, are specialist predators of other fish species as well as their own. Among the amphibians and reptiles, the grass snake is probably the most important predator of well-grown fishes; gudgeon up to 12cm in length are recorded as having been taken. The eggs and small larvae of some species are also preyed upon by large invertebrates, such as dragonfly nymphs, water beetles, water spiders and crayfish.

Birds such as the divers and grebes, grey heron, sawbill ducks, osprey and kingfisher are reliant on freshwater fishes as major components of their diet. Two species, cormorant and common tern, which were formerly regarded as being mainly dependent on marine fishes, have in recent decades increasingly colonized inland waters or have widened their foraging ranges to include freshwater areas. Consequently, cormorants are now regarded as serious pests at some freshwater fisheries, but the increasing numbers of common terns in river valleys seem to have had no discernible effect on fisheries. Sawbill ducks are regarded as pests in some areas with salmonid fisheries and red-breasted merganser and goosander have been controlled by both licensed and illegal shooting, although the effectiveness of this control has been questioned. The return of the osprey to Britain as a breeding species has been welcomed even though they prey mainly on freshwater fishes in areas important for angling. Bitterns, now regrettably rare in Britain, are reported to feed mainly on eels. Many other species of birds are reported to include freshwater fishes as part of a wide range of food items. These include several species of waders, ducks, owls and raptors as well as water rail, coot, mute swan and dipper.

Two mammals regularly prey on freshwater fishes in Britain: American mink and otter. Both species are opportunistic predators, able to catch and eat anything from invertebrates to other mammals, but freshwater fishes are an important part of their diets, especially the otter. American mink, an introduced species, is increasingly regarded as an undesirable pest, whereas otters are being re-introduced to areas from which they were lost during the twentieth century. Red foxes are omnivorous and recorded as including fishes in their diet, but probably as the result of scavenging. Water shrews and, exceptionally, water voles also feed on freshwater fishes. Seals, both common and grey, feed mainly on fishes and inevitably prey on anadromous species such as salmon and sea trout at sea and in the mouths of tidal rivers, but studies suggest that their main fish prey comprises wholly marine species.

Cultural consciousness of freshwater fishes in Britain

Fishes in naming the landscape

Evidence for awareness of freshwater fishes can be sought in how our ancestors perceived, used and named our landscape. It is widely accepted that Ely takes its name from the Anglo-Saxon Elge or Elige meaning 'eel district' signifying the particular importance of this area for eels. Eel Pie Island in the Thames at Twickenham, Middlesex, was certainly associated with a local eel fishery. Fishbourne, Sussex; Fishburn, Co. Durham; and Fishlake, Yorkshire probably refer to the importance of freshwater fishes in local watercourses. The Fishponds area of Bristol takes its name from a hamlet near quarry pits that were disused and became water-filled in the early seventeenth century but which have long since been filled in and built over. Skelly Nab, a promontory on the northern shore of Ullswater in Cumbria, is the only known spawning ground for schelly in this lake; skelly is a spelling variant of schelly. However, place-names in English that are apparently associated with plants, animals and habitats can be deceptive. For example, the Shadwell district of East London is said by Richard Fitter in *London's Natural History* to be associated with shad, which were fished commercially in the Thames, but place name experts suggest the meaning 'shallow well', 'boundary well' or more probably 'St Chad's well'.

The Celtic languages of Welsh, Scottish Gaelic, Cornish and Manx include species in place names, but these can be obscured in any anglicization. The Gaelic name for trout (breac) is found in some Scottish placenames, such as Loch nan Breac, of which there are several examples. The Norse word for salmon (lax) is also found in the names of several burns and rivers in Scotland, for example Laxford in Sutherland, Laxadale on Lewis, and also at Laxey on the Isle of Man. Why Loch Spiggie in Shetland is apparently named from the Scandinavian word for stickleback (spigg) is somewhat baffling.

Field names also contain pointers to the history of the landscape, 'Fishpond' being the most common name obviously related to freshwater fishes. The golden age of fishponds in Britain was in the Middle Ages and although some field names from this period, such as 'Fishpond' or 'The Stews', may survive to this day, more can be found in pre-enclosure estate maps from the seventeenth and early eighteenth centuries.

Fishes as food – a historical perspective

Freshwater fishes have formed part of the human diet throughout the post-glacial period for roughly the last 10,000 years, ever since the earliest Upper Palaeolithic colonists spread through Britain. Evidence from archaeological excavations and from cave decorations and ornamented artefacts in nearby continental Europe, shows that salmonids, pike, perch, roach and eels were used by these people. The exploitation of wild stocks of freshwater fishes continued through the Mesolithic and Neolithic periods and the Bronze and Iron Ages. Tantalizing fragments of information about the apparent use of freshwater fishes for food are scattered through the archaeological literature. These include sturgeon from a Mesolithic site at Morton, Fife; pike from Mesolithic layers in two caves in Derbyshire; trout and eels at the Neolithic village of Skara Brae, Orkney; and salmon or trout and roach from an Iron Age settlement at Meare Village East in the Somerset Levels. Detailed excavations of Middle Neolithic and Bronze Age sites at Runnymede beside the River Thames in Berkshire, and a Middle Iron Age site at Haddenham in the Cambridgeshire fens, produced evidence of very limited use of freshwater fishes for food. It revealed that pike and cyprinids were used at both sites and salmon at Runnymede only. These studies have led to the tentative conclusion that freshwater fishes formed an insignificant part of the diet of prehistoric farming communities in southern Britain.

The Roman period provides what may be the earliest evidence of the culture of freshwater fishes in Britain. Whether this was merely the holding of wild-caught specimens in ponds and tanks or active breeding in captivity is debatable. Thereafter, artificial water bodies, such as moats, dams and ponds, were created, but their purposes appear to have been many and various, though keeping or rearing fishes would have been an important one. Ponds (or 'stews') for the culture of freshwater fishes for food are noted in the Domesday survey of 1086 and are well documented from the twelfth century. Records of thefts of fishes from manorial and monastic ponds are a particularly good source of information on what species were kept and in what quantities, and in some cases their commercial value. From the Anglo-Saxon period until at least the fourteenth century, fish weirs in tidal estuaries to catch both marine and freshwater species were also important. It has been estimated that there were 500 such weirs around the coast of England alone.

There is other good evidence of commercial fisheries exploiting wild stocks in the Anglo-Saxon period. In the late 970s, for example, 20 fishermen 'rendered' 60,000 eels annually to the brethren of Ramsey Abbey, Cambridgeshire. This form of taxation of fisheries was commonplace in the eleventh century and the Domesday survey records taxable fisheries in most English counties, mainly with respect to eels, but also salmon and lampreys. Evidence of these fisheries takes several forms, such as the payment of dues for seine- and drag-nets on the Thames at Hampton, Middlesex, and 'renders' (rents or taxes) paid in eels from watermills and fish weirs in many counties.

The productivity of the Domesday fisheries appears to vary greatly when expressed as renders paid either in eels or in money. For example, renders of 24,000 eels from Stuntney and 27,150 eels from Doddington, both in Cambridgeshire, suggest that if these quantities of eels were being levied as taxes or rent, vast numbers of eels were being fished commercially from the Cambridgeshire fens. Elsewhere in England, renders of only 500 or 1000 eels were more common. Eels fulfilled many of the functions of currency in the Fenland area at this time. The fisheries of mediaeval Fenland were not limited to eels. Writing about the Fens in around 1125, William of Malmesbury listed also pickerel (pike), perch, roach, burbot, lamprey and sometimes sturgeon, all of which were presumably eaten.

Excavations of the kitchens at the site of the Benedictine abbey at Eynsham, not far from the river Thames near Oxford, showed that freshwater fishes formed a significant part of the diet, something more often found in continental Europe than in Britain. The finds, which span the eleventh to the sixteenth centuries, include pike, eel, dace, perch and unidentified cyprinids, as well as sturgeon and stickleback, although whether the last were actually eaten is unclear. There are also many marine species, particularly from later periods, possibly brought in to the Abbey in a preserved form (for example salted, pickled or smoked) or even as fresh fish – the Abbey was probably less than two days from tidal waters by boat and road. The finds indicate that, over time, the pike eaten at Eynsham declined in size. This was possibly due to increased exploitation, but more probably due to the catches in earlier periods being wild-caught and, in later periods, being from fish-ponds known to have been constructed in the thirteenth century.

Freshwater fishes probably provided a useful source of protein for people in inland, rural Britain until at least the seventeenth century. This would have been especially important in the lean months at the end of winter and in spring when supplies of stored pulses and preserved meat were exhausted. Fishes could also be eaten fresh or preserved, but preservation was probably used mainly for marine species. The tradition of eating fish on Fridays, which has its origins in pre-Reformation Christian dietary observance, was also important in determining the demand for fishes of any type in the diet. Although this tradition persists to this day in Britain among a minority, it has now become almost a folk memory. Freshwater fishes were almost entirely replaced by marine species as soon as rapid transport – first by road, then by rail, and now again by road as well as by air – became available. However, there is good evidence to suggest that, as early as the fourteenth century, fresh marine fishes were transported to inland cities and towns though perhaps not to remote country areas. As the population increased and sea fishing techniques and transport improved, so the demand for freshwater fishes became more localized.

Commercial fisheries based on freshwater species, other than salmonids and eels, declined progressively, although remaining in several areas into the twentieth century (Plate 6). For example, net fisheries for perch, pike, eels, trout and Arctic charr existed on Windermere until 1921, but were then closed down due to over-fishing. This recent history of commercial fishing in Windermere prompted the first full-time Director of the Freshwater Biological Association, Barton Worthington, to promote commercial trapping of perch between 1941 and 1947 as a contribution to the war effort. Sardine-sized perch from Windermere were sent to Leeds, where they were cooked in tomato and Yorkshire relish, canned and marketed as 'Perchines'. During the lean years of the Second World War and consequent food rationing, they would have made an interesting addition to the national diet!

Twenty-first century tastes in Britain are more conservative, and salmon and trout (mainly rainbow) predominate in the non-farmed commercial fisheries and angling catches for food. Wild salmon, and sea/brown trout fisheries are protected legally and closely guarded by those with rights over them. A few commercial eel fisheries survive, but mainly for exporting to continental Europe and the Far East, although a limited market for jellied and smoked eels still exists in Britain. More adventurous anglers may eat some of the species they catch, but in Britain we seem to have largely turned our backs on the considerable culinary quality of pike, perch, tench, barbel, carp, grayling, zander and even river lamprey; species that are still eaten in continental Europe.

Fishes and fishing in art

Fishes feature as decorative motifs from the Roman period onwards in Britain. Fishes and fishing acquired symbolic meaning early on in the spread of Christianity within the Roman Empire, and the simple outline of a fish remains a Christian symbol to this day. Some notable examples of such use of fishes in decoration are among the Anglo-Saxon grave goods from ship burials at Sutton Hoo, Suffolk, in Norman architectural and monumental decoration such as at Kilpeck Church, Herefordshire, and in ecclesiastical wood-carvings, as in Exeter Cathedral, Devon. In northern Scotland, eighth-century carved Pictish stones bear outlines of life-sized salmon, often recognizably male, but their purpose is unknown. The boundaries between religious symbolism and aesthetic art become clearer in the late Middle Ages for it is then that depictions of fishes or fishing begin to appear in paintings and illustrations. One of the earliest examples is the woodcut frontispiece to *The Treatyse of Fysshynge wyth an Angle*, attributed to Dame Juliana Berner and published by Wynken de Worde in 1496. This depicts an angler with rod, line and float hauling in a fish to the bank of a river. In the middle background is what appears to be a tank or barrel containing other fishes in water.

The first British artist to specialize in sporting subjects was probably Francis Barlow (d. 1704). An engraving by Wenceslaus Holler of one of Barlow's paintings, *River Fishing*, was published in 1674. By the mid-eighteenth century angling was clearly acceptable, even fashionable, in Britain as a suitable subject for portraits and conversation pieces (informal group portraits). There are examples by important artists such as William Hogarth, Johann Zoffany, Arthur Devis,

George Morland and Sir Henry Raeburn. It is interesting to observe that many of these conversation pieces include women and children as the anglers. Fishes and fishing are also frequent decorative subjects on porcelain produced in the mid-eighteenth century.

At the beginning of the eighteenth century the depiction of anglers in landscape painting (for example by Thomas Ross), probably drew on influences from Dutch artists, but from the 1770s onwards artists such as Thomas Hearne, John Crome and J. M. W. Turner developed a uniquely British feel for realistic landscapes with figures, including anglers. Equally important are the wood engravings of Thomas Bewick that capture many aspects of country life, such as fishing with rod-and-line, from a truly rural perspective.

By the nineteenth century angling was fashionable and action portraits of individuals or groups of gentlemen fishing were in vogue throughout most of that century. The artistic merit of most of these paintings is not great. Of equally dubious quality are the often whimsical angling pictures of children or rustics that were fashionable during the reigns of Queen Victoria and King Edward VII. Members of the Rolfe family were prolific in producing remarkably lifeless paintings of fishing scenes during this period.

The fishes themselves also feature as the subject of paintings. The earliest examples follow the style of seventeenth-century Dutch still-life painters. Paintings of dead fishes abound from the eighteenth and nineteenth centuries but, by the end of the 1800s, action portraits of fishes rising or being hooked or landed, by artists such as A. Rowland Knight, had become popular. The quality of illustration in some of the nineteenth-century taxonomic guides to fishes, such as those reproduced in this book, needs no further commendation. This tradition of scientific accuracy has been maintained up to the present day by illustrators such as Stuart Carter and Martin Knowelden.

There is still a flourishing market in Britain for realistic, contemporary paintings, sculptures and ceramics of fishes, with established artists such as Robin Armstrong, Neil Dalrymple, David Hughes, Charles Jardine, Terence Lambert, Rodger McPhail and Richard Smith active at the present time. Pleasing artistic interpretation of their fishy subjects is a feature of the work of Leonard Appelbee and Colin Paynton. Exhibitions of the Society of Wildlife Artists regularly include works by members who specialize in fishes.

On a lighter note, it is worth noting that, since the late eighteenth century, fishing and anglers have been the subjects of humorous drawings, some of which were politically satirical. *Punch* featured cartoons involving angling, often dwelling on the apparent futility of the activity. H. M. Bateman, the prolific angling illustrator and cartoonist, was a frequent contributor to that magazine during the 1930s and '40s. More recently, Norman Thelwell, best known for cartoons featuring small girls on overweight ponies, has produced some notable cartoons on the theme of fishing, and the widely syndicated American cartoonist, Mark Parisi, usually depicts the perils of fishermen from the fishes' perspective.

Fishes as decorative trophies

Yellowing glass cases with lifelike casts of heavyweight fishes used to be a feature of certain types of country pub and small hotel the length and breadth of Britain. Many probably found their way to these resting places via local salerooms, as the contents of gun- and rod-rooms of Victorian and Edwardian country houses were dispersed during the twentieth century. The fashion for casts of specimen fishes seems to have declined in Britain in recent decades, but this form of recording notable catches still appeals to some anglers.

Fishes and fishing in literature

'I love any discourse of rivers, and fish and fishing' ... wrote Izaak Walton in *The Compleat Angler*.

As with any special interest activity, whether it be fishing, philately, football or photography, there is a wealth of associated literature. Most of the books relating to freshwater fishes in Britain are about how and where to fish successfully, with many books and pamphlets devoted to fishing for a single species. This technical literature of angling is large and diverse, and to review it could occupy a book of its own. Such is the international nature of angling, British bookshops specializing in fishing literature often include books by authors from abroad, especially from

North America where there are several regional associations of angling writers.

Among the books on how and where to fish there is also an interesting genre of fishing memoirs and reminiscences. One of the earliest and most famous books of this kind is Izaak Walton's *The Compleat Angler, or, The Contemplative Man's Recreation*, published in 1653 as '*a discourse of Fish and Fishing, not unworthy the Perusal of most Anglers*'. The subtitle to Walton's book reflects the particularly distinctive quality of many of such works. Their authors take pleasure in all aspects of fishing: they appreciate their surroundings and they express a sense of competition with 'nature', be it the fishes, the location or the weather (Plate 7). The appeal of fishing for many anglers is summed up by the title of Jeremy Paxman's eclectic anthology, *Fish, Fishing and the Meaning of Life*, or more simply by the jazz singer, writer and art expert, George Melly, in the ambiguous title of his characteristically vibrant angling memoir, *Hooked!* As with the technical literature on angling, many volumes of memoirs by North American authors are also available in Britain.

Because so many people in Britain go fishing, and have done so for recreation for at least 500 years, images of fishes and fishing occur frequently in our literature. They appear in poetry, literary prose, crime, adventure and travel writing as well as in works for children. They are to be found as metaphor for aspects of the human condition as well as in descriptive pieces. These images often draw on the beauty of an individual fish; on the challenge presented by the angler's skill pitted against wild nature; and on the 'spirit of place'. Two examples (among thousands) that characterize the range of this writing have a distinctive British dimension. John Buchan's *John Macnab* is a classic story of adventures in the Highlands on the theme of fishing for salmon and stalking red deer. Ted Hughes' atmospheric poems in *River* include a description of a run of sea trout. Fishes and fishing also appear in humorous writing, such as in the nonsense verses of Edward Lear and in Jerome K. Jerome's memorable account of exaggeration by anglers in *Three men in a boat*.

A further genre that should be considered is general natural history writing. Here it is surprising how few publications treat freshwater fishes in any detail and many do not mention them at all. Although some of the regional natural histories published in the late nineteenth and early twentieth centuries cover freshwater fishes thoroughly, few of the natural history sections of the Victoria County Histories include more than a simple list of species. More recently, even serious books on regional natural history, such as several of the titles in the New Naturalist series, give few details of the occurrence of freshwater fishes. This tends to confirm a rather perplexing impression that for most 'naturalists' freshwater fishes are normally out of sight and out of mind!

Study and surveys of freshwater fishes in Britain

Research and surveys

Much has been written since at least the seventeenth century about the general biology of some of our fishes, in particular about how to catch the popular angling species. The classic nineteenth-century treatises draw on a mixture of careful observation and folklore, which make entertaining reading. Our present-day knowledge of the ecology, biology and distribution of freshwater fishes has accumulated gradually and progressively throughout the twentieth century. For example, the migratory life history of the European eel was not described until the 1920s. Ecological research on individual species of freshwater fishes has been going on since at least the 1930s in Britain (Plates 8 and 9), including at several universities; the Freshwater Biological Association; Fisheries Research Services; Centre for Environment, Fisheries and Aquaculture Science; the Centre for Ecology and Hydrology; and commercial organizations and consultancies. The individual species accounts accompanying the distribution maps draw on the cumulative results of this research.

Identification publications

Among the many books on British freshwater fishes, there are several important guides to the species, many of which are illustrated. It is impractical to cover all such publications, but some of the more notable examples are listed chronologically in Appendix 1.

The distinction and erudition of many of the authors of these works is noteworthy. The earliest in the seventeenth century, Francis Willughby, a founder Fellow of the Royal Society, was a friend of Samuel Pepys and the patron of John Ray, considered the father of British natural history. John Williamson's mid-eighteenth-century pocket companion covers in great detail methods of fishing and fish husbandry, including how to maintain and stock artificial fish ponds. At the start of the nineteenth century, Edward Donovan was one of the most prolific writers on natural history at a time when it was the most fashionable and popular of the sciences in Britain, and he was followed by William Yarrell, Jonathan Couch, William Houghton and Francis Day in producing comprehensive accounts of British fishes. In the twentieth century, relatively few identification guides to freshwater fishes were published up to the 1960s, but during the last 40 years several authors have published new guides. Alwyne Wheeler, Peter Maitland and Malcolm Greenhalgh among others have written well-researched texts on British freshwater fishes and their identification, thereby re-establishing the previous strong tradition of authoritative works.

Further reading:

The importance of freshwater fishes – Anon, 2003.

Fishes and other wildlife – Corbet & Harris, 1991; Maitland & Campbell, 1992.

Cultural consciousness – Ayres *et al.*, 2003; Clarke, 1998; Coles, 1987; Darby, 1940; Farrar, 1998; Gelling, 1984; Le Cren, 2001; Morris, 1989; Quinn, 1991; Rackham, 1986; Serjeantson *et al.*, 1994; Simmons *et al.*, 1981; Steane, 1971; Wilson, 1973.

Study and surveys – see Appendix 1

2 The distribution of freshwater fishes in Britain

The origins of the native British fishes

During the last Ice Age, Britain was largely covered by the ice sheets that extended over much of northern Europe. The southern part of what is now the British mainland, although not beneath the ice, was a region of freezing climate and a landscape of permanently frozen tundra. Throughout this period freshwater fishes were absent from Britain, as they were from much of the rest of northern Europe, excluded by these harsh climatic conditions.

Around 12,000 years ago, the climate started to warm and the ice began to retreat northwards. Fishes had survived in continental Europe in various ice-free refuges, including major ones in the Iberian peninsula, the Po basin in northern Italy, the Danube basin (the largest glacial refuge), the Balkans and Greece. Slowly, fishes from these refuges began to spread northwards and westwards, extending their range by colonizing waters newly created by the melting ice. Recent evidence suggests that cold-adapted species may also have survived close to the glacial margin in more northerly glacial refuges, possibly even in Britain. Lakes were formed by the huge volumes of melt water, and anadromous forms of species such as the whitefish and Arctic charr, tolerant of low water temperatures, were probably the first fishes to colonize these cold, new lakes. Later, as the melt water drained away, these populations were left isolated, and can still be found today in the Lake District, Scotland and Wales.

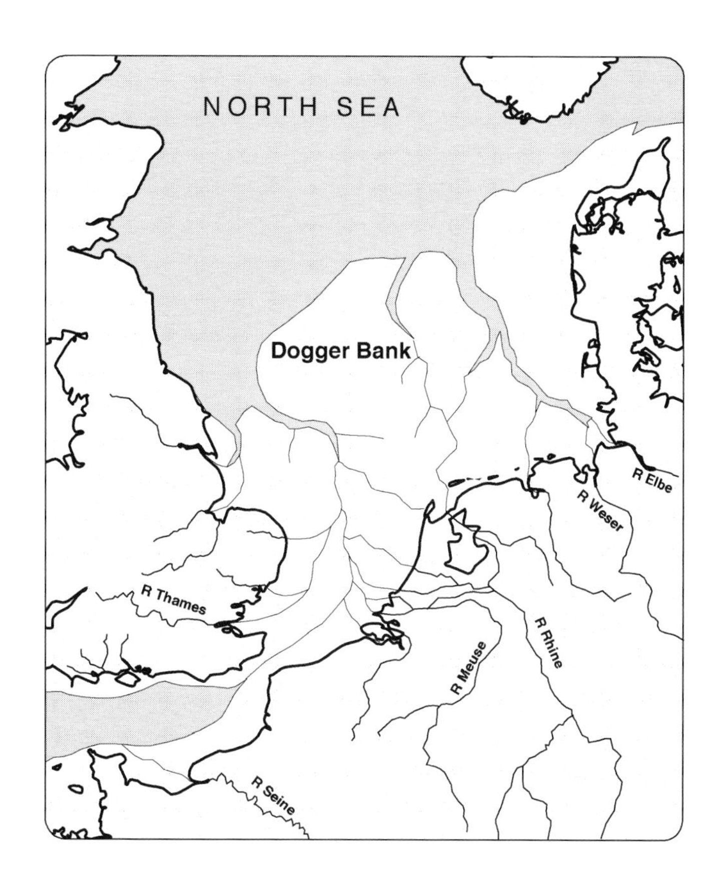

Figure 1
Britain showing land bridge and river network 10,000 years ago. The present-day coastline is shown in bold.

The melting ice produced an extensive network of rivers and lakes, allowing other species of fishes to migrate into these areas as temperatures rose. Many freshwater species expanded their range across Europe as the ice retreated, eventually reaching Britain from the rivers of western Europe such as the Rhine, of which the Thames was then a tributary.

At this time, Britain was still joined to the rest of Europe by a land bridge in the south-east (Fig.1). Freshwater fishes were able to cross this land bridge via the network of rivers from what is now continental Europe into the rivers of Britain, although this migration could not continue to Ireland which by then had already become separated from Britain by the rising sea levels. Several species that crossed into Britain, notably the barbel, spined loach and burbot, remained in these eastern river catchments, never naturally extending their range to the rest of the country. By 6500 years ago, perhaps as early as 9000 years ago, the melting ice had raised sea levels sufficiently to open the Straits of Dover, drowning the land bridge and breaking the freshwater link with the rest of Europe. The isolation of Britain prevented any further migration by species incapable of tolerating saline conditions, which explains why species such as the schneider *Alburnoides bipunctatus* Bloch and the nase *Chondrostoma nasus* Linnaeus are present in mainland Europe but absent from Britain. However, species capable of tolerating fully saline conditions, including salmon, sea lamprey, eel and trout, would have continued to colonize the new island.

Natural distribution of fishes

The natural distribution of fishes in Britain has been determined by their post-glacial dispersal, the biological and ecological requirements of each species, and their interactions with other species in their environment. At the broadest scale, the natural pattern of distribution of fishes across Britain is determined by the prevailing climate and geography. Across broad geographic regions, the presence of given species within specific catchments reflects local topography and the accessibility of the catchment. Within each catchment, the distribution of fishes is governed by increasingly local effects, such as water temperature, flow velocity, productivity, substrate type, gradient and other factors. These factors affect species distribution, or alter the structure of the fish community and abundance of the various species present on a local scale, but have limited influence on the broader pattern of species distributions on a national scale.

Mechanisms for natural dispersal of fishes

Once present in a river system, fishes spread through the network of tributaries (except where there are impassable barriers such as waterfalls), occupying areas where the habitat is suitable. Flooding causes rivers to overflow their banks, often inundating large areas with floodwater, and can allow fishes and their spawn to be transferred to other watercourses and to still waters, where they remain after the floodwaters subside. Before the nineteenth century, wetlands were much more extensive than today and flooding was a more common event, which assisted natural dispersal. Fishes and their spawn can also be displaced in a catchment by river flow, increasing the range of a species downstream within the system. Anadromous and catadramous species, which spend a part of their life cycle in the open sea, are able to spread to other catchments on their return to fresh water by entering different river systems to those from which they originated. The adhesive eggs of some fish species can be transferred from one water body to another on the fur of mammals or the plumage and legs of birds. Eels are alone in being able to travel overland in wet conditions, often reaching isolated water bodies, giving the species a very wide distribution.

Environmental factors influencing the distribution of fishes

Every fish species, and each stage in the life history of every species, has its own set of preferred habitat requirements, reflecting its particular biology and behaviour. Those species with a wide tolerance of environmental conditions, such as eel, can be found in diverse habitats ranging from the headwaters of upland rivers to nutrient rich lowland lakes. They are able to survive in a wide range of habitats and therefore tend to have a more widespread natural distribution than species with narrower habitat requirements, such as grayling. Salmon, which require good water quality, cool water and a stony substrate for spawning are restricted to rivers that offer these conditions but are absent from catchments that do not, as is the case in parts of East Anglia and the Midlands.

The prevailing climatic conditions affect the distribution of fishes throughout Britain. Some species require relatively high water temperatures in order to spawn successfully – tench, for example, can spawn only when springtime water temperatures reach 18°C. Cooler springtime temperatures in the north of Britain may prevent these fishes from spawning successfully in some years, putting these more northerly areas on the edge of their geographic range. Similarly, cold-water species such as the salmon, comfortable in the temperatures offered in British latitudes and as far south as Portugal and northern Spain, are absent further south, where water temperatures are too high to allow them to reproduce successfully.

Variations in water temperature between years have a profound effect upon spawning success in coarse fish species. When water temperatures are relatively high at the time when fish fry are emerging, and for the first few weeks afterwards, the fry experience better survival rates than in cooler years. These strong year classes can be followed through in subsequent years, when they continue to dominate the population structure.

The natural variations in river flow and still water levels brought about by variable rainfall patterns also affect fish populations. High flow rates displace fry, increasing the mortality of juvenile fishes, and flooding can leave fishes and fry stranded. During periods of drought, low water levels can impede the passage of fishes and create high water temperatures, lethal to temperature-sensitive species such as salmon and trout. When lakes or rivers lose water, fishes become more crowded and confined in a smaller volume of water where they become increasingly stressed. As the water temperature increases, the concentration of dissolved oxygen in the water falls, further stressing the fishes. In extreme cases, evaporation may also concentrate waste products in the water, such as ammonia and carbon dioxide, which can also stress or kill fishes. If drought persists, rivers can run dry, and ponds and even lakes empty, killing all fish species.

The habitats provided by different fresh waters are important in determining which species are present. Each water body, whether of flowing or still water, provides a range of environments influenced by the topography and geology of the catchment. Catchment geology affects nutrient availability: catchments with acidic geology, such as granite, have much lower levels of nutrients than are present in catchments with neutral or calcareous geology, such as chalk. The availability of nutrients influences the biological productivity of a water body – a nutrient rich lake is much more productive, and therefore able to support a greater population of fishes than a nutrient poor lake.

This difference in nutrient status is reflected in the typical population of each type of water: trout and perhaps Arctic charr or whitefish are found in the cool, oxygen rich but unproductive habitat offered by a tarn, to which they are well adapted but in which water temperatures are too low to allow most cyprinid species to spawn. A lowland lake provides an ideal habitat for bream, roach, tench or other cyprinids adapted to this more productive environment. Here, higher water temperatures and lower levels of dissolved oxygen exclude cooler water species such as trout, which are unable to tolerate these conditions. The topography of the catchment affects the type and distribution of available habitats, and therefore affects the species present and their distribution: the community in the fast-flowing headwaters of a river (Plate 10) is very different from that of the sluggish lowland reaches.

River habitats

The diverse nature of river habitats enables rivers to support a greater variety of fish species than the more uniform habitats found in still waters and canals. The changing morphology typically offered by a river, as it flows from its headwaters to its mouth, produces a succession of habitats, each supporting a different community of fishes. This changing habitat is largely dependent upon the gradient and the width of the watercourse, which in turn affects other physical characteristics including the velocity of the current, the nature of the substrate and the levels of dissolved oxygen. These physical factors are also important in determining the abundance and composition of the invertebrate and plant communities of the river. The concept of changing habitat influencing the structure of the fish community is exemplified by

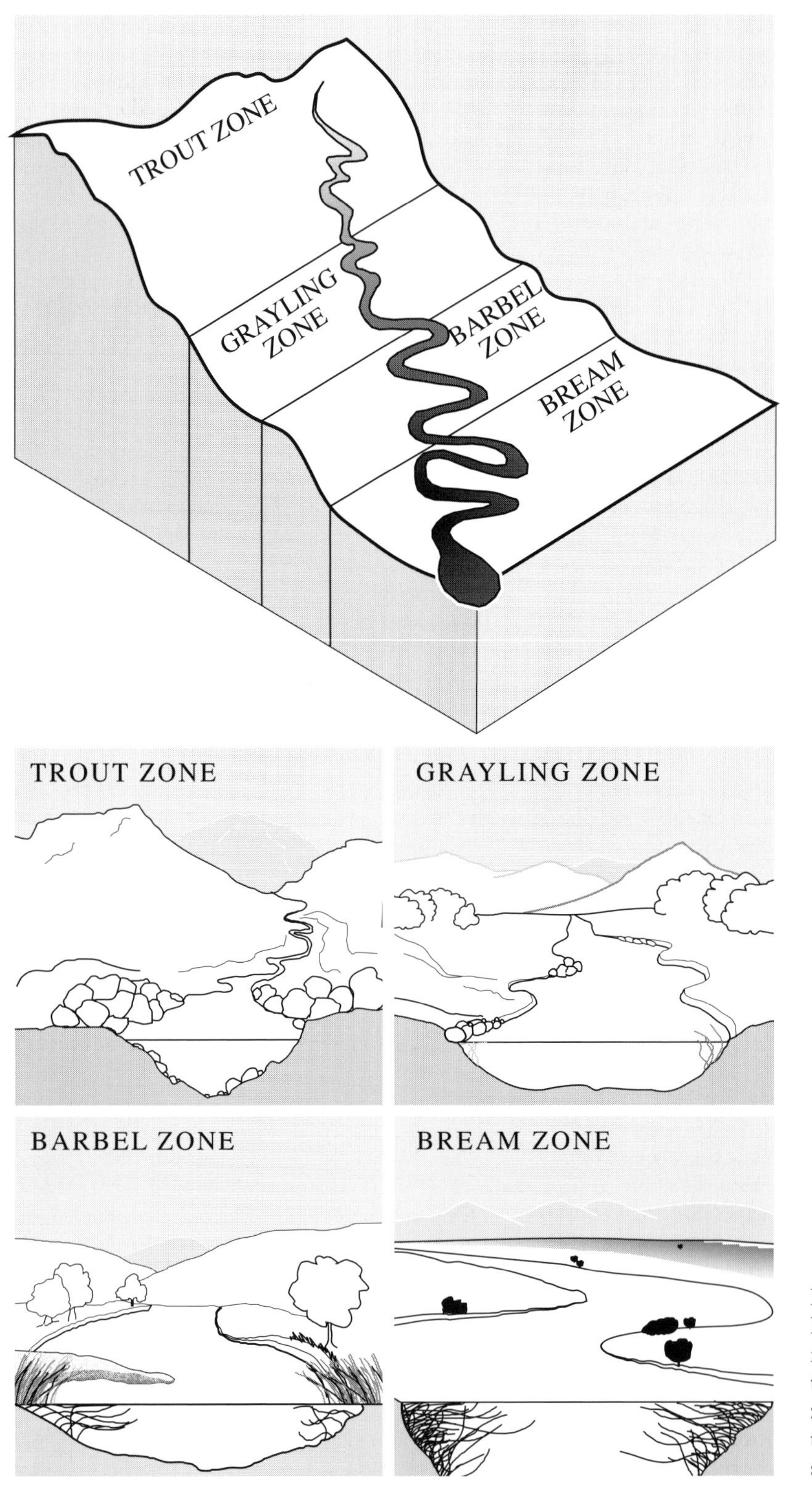

Figure 2
Diagrammatic representations of the main river zones named for the dominant fish species present.

the classic river zonation devised by Huet (1949), who divided rivers into zones named for the dominant fish species present.

The **trout zone** (Fig. 2; Plate 11) represents the headwaters of the catchment, where the steep gradient, high energy, eroding stream flows through a narrow valley, with an unstable boulder and cobble substrate. There are often waterfalls and cascades, which help to oxygenate the water. Higher plants are often absent due to the fast current and lack of nutrients in this zone, although aquatic mosses and algae may be present. Much of the food available for fishes in the stream is from terrestrial insects and plant material falling into the watercourse. The difficult environment in this zone means that there are few species of fishes, typically trout and bullhead.

The **grayling zone** (Fig. 2) is next in succession, where the river has a gentler gradient, greater width and depth, and a slower current. This gives rise to a substrate consisting of sand and gravel, as well as cobble and boulders, with a succession of faster flowing riffles separated by pools. Mosses and aquatic plants are likely to be found in this zone, which has more nutrients than the trout zone, derived from the surrounding catchment. Fishes typical of this zone are grayling, salmon, trout, eel and minnow.

Further downstream is the **barbel zone** (Fig. 2; Plate 12) where the decreasing gradient continues to reduce the current and the input from tributaries increases the discharge of the river. The channel is wider, the sediment load greater and the substrate finer than in the zones upstream. The channel meanders and both marginal and instream vegetation develop, often in thick stands. The river has a sequence of pools and riffles, and is typified by a more diverse fish community including barbel, chub, dace, bleak, roach, pike and perch.

The **bream zone** (Fig. 2) is the lowest reach in the freshwater river. It has the least gradient, greatest flow and highest levels of turbidity and nutrients. The substrate is finer, with sand, gravel and silt deposited by the slow-flowing river. Dissolved oxygen levels may still be fairly high at the surface, where the water is in contact with the atmosphere, but may be lower near the river bed. Water near the bed of the river does not readily mix with the oxygenated upper layers and bacterial action in the bottom sediments removes oxygen from the water. The river meanders across a wide floodplain, and can contain a large number of fish species including bream, gudgeon, carp, roach and other cyprinids, with migratory fishes such as salmon, trout, lampreys and eels passing through. The environment offered by man-made watercourses, such as canals and drains, is similar to that found in the bream zone.

Below the bream zone the river mixes with the sea, forming the estuary. The freshwater flow may or may not mix with sea water, depending on both the nature of each individual estuary and the state of the tide. Here, the variable salinity resulting from the ebb and flow of tides provides conditions that not all fishes can tolerate. Estuaries are rich in food sources, and marine species including flounder, grey mullets and sprat move from the sea to use estuaries as nursery areas. The substrate is composed of silt and mud and the water is warmer, becoming increasingly saline towards the sea. Typical freshwater species present include perch and roach, which are often found in the upper, less saline reaches, and migratory fish, on their passage to and from fresh water.

In practice, rather than an abrupt change from one zone to another, gradual changes in the species composition are found, as the river habitat changes and the distribution ranges of the species overlap. However, not all rivers follow this general pattern. The chalk streams of southern England do not show this zonation, and very short rivers, as are commonly found in the west Highlands of Scotland, may be composed entirely of one zone. In other rivers the zones may repeat themselves, or be omitted, variations which are dependent on the local topography. River engineering, for power generation, navigation and flood defence, can modify the natural patterns of occurrence of these zones.

Still water habitats

The habitats offered by lakes, reservoirs, ponds and other standing waters differ from those of rivers principally in that there is no constant direction of flow. Within lakes there are a number of different habitats, as in rivers. At the lake edge, sunlight may penetrate to the lake bed, promoting plant growth. Fishes may spawn on the vegetation in this weedy area, the littoral zone, and feed upon the large numbers of invertebrates that the plant growth supports. The open water, or pelagic zone, offers a different habitat. There are no higher plants in the open water, and the fishes of this zone, including Arctic charr and whitefish in upland lakes, and roach and rudd in lowland lakes, feed mostly on zooplankton. The deepest zone of lakes, where sunlight does not penetrate, is the profundal zone. Light penetration is dependent on the amount of suspended solids and algae in the water column, which reduce the depth to which light reaches. The absence of light prevents the growth of plant life, however, the silt and mud of the lake bed contain chironomid larvae, molluscs and other burrowing invertebrates, which provide a source of food for bottom feeding fishes such as bream and tench.

The most important factor controlling the fish species of still water habitats is their productivity, or trophic status.

Nutrient poor (oligotrophic) waters, typically draining upland acidic catchments, are cold, well oxygenated and clear, but contain only low levels of the nutrients required for aquatic plant and algal growth (Plate 13). Consequently, these lakes are able to support only a small aquatic plant community, which in turn supports a sparse invertebrate community. This cold, low productivity environment provides a habitat suitable for trout, eel, and, where accessible from the sea, salmon. Such lakes may contain relict populations of charr or whitefish, present since the end of the Ice Age. Examples of oligotrophic waters include many of the lochs in the Scottish Highlands, and many upland reservoirs and mountain tarns.

Nutrient rich (eutrophic) lakes are generally shallow, fertile, lowland waters with a high nutrient input that promotes extensive growth of higher plants and algae. This abundant growth supports a large invertebrate community, providing plentiful food supplies for fishes. These lakes are usually turbid, and may suffer from low levels of oxygen, particularly during the summer months. As a result, fishes with a high oxygen requirement, such as salmonids, are often absent from this type of water. Species adapted to these conditions, including carp, roach, rudd, perch, tench and bream, are often present in large numbers, supported by the high productivity of the habitat. Eutrophic waters include the majority of farm ponds and urban lakes.

Water bodies not falling into either of these categories are of moderate (mesotrophic) nutrient status, and provide an intermediate habitat for fish. They receive a moderate amount of nutrient input, which produces more plant growth than in oligotrophic waters, but they are less enriched, and therefore less productive, than eutrophic waters. They may contain trout, which are likely to exhibit better growth than trout in oligotrophic waters, or other species generally associated with low nutrient status waters, as well as cyprinids and other fish, including roach, rudd, perch and pike.

Human influences on the distribution of fishes

The distribution of British fishes has been greatly modified by human activities, both directly by affecting the fishes themselves and indirectly by affecting their environment. Sometimes, human intervention has been with the specific aim of modifying the distribution of species, for example by stocking (although the results are not always those expected), but in most cases the effects on fish populations have been entirely incidental. There is no *intent* to affect fish populations by, for example, damming a river to store water; straightening a channel to improve flood protection; abstracting water for domestic, industrial and agricultural use; fertilizing arable land to improve crop yields; or discharging waste water. Nevertheless these activities, and many others, may significantly affect the distribution and abundance of fishes.

The ability of humans to change their environment (and that of all other species) has grown steadily for at least the last 3000 years, but the rate of change has been very rapid in the last 50 years. Management of the topography of all types of wetlands and the creation of new water bodies have inevitably had a profound effect on the distribution and abundance of freshwater fishes. The Romans began draining wetland areas such as parts of the Fens, but at the same time created a new water body, the Car Dyke, a 145km long canal between Lincoln and Waterbeach, near Cambridge. The Norfolk Broads developed from large-scale, mediaeval peat diggings in areas of fens in the valleys of the Rivers Ant, Bure, Flegg, Thurne, Waveney and Yare. The main period of canal building in Britain was in the eighteenth century, linking widely separated river catchments for the first time. Thus it was possible by the nineteenth century to travel by water from the

Thames to the catchments of the Great Ouse, Severn, Trent, Humber and Mersey, most of which system is still open for the movement of aquatic wildlife (Fig. 3; Plate 14).

The effects on freshwater fish of most of these types of changes in water bodies are almost unknown, although there are a few studies of fish populations where water levels have been raised such as at Loch Tummel, Perthshire; Loch Garry, Inverness-shire; and Cow Green Reservoir, Co. Durham.

Water quality and pollution

All fishes require adequate water quality if they are to survive. If water quality falls below the minimum required by a species, fishes will either be killed or forced into other areas. Persistent, poor water quality can exclude species from entire catchments where they were once present, and a single pollution incident can kill all the fishes in a river for many kilometres downstream, and these effects may persist for years. Water quality must be maintained if populations are to thrive.

Individual species have different water quality requirements, with salmonids requiring better water quality than cyprinids. These differences are recognized in the standards set by the European Commission in 78/569/EEC Directive on the quality of fresh waters needing protection or improvement in order to support fish life, more commonly known as the 'Freshwater Fish Directive', which aims to ensure that in those waters where fishes are present, water quality is adequate to support them. Table 1 (p. 26) shows the recommended (guideline) and maximum acceptable (imperative) values for a range of water quality parameters (with the exception of the values for dissolved oxygen, which are minimum levels) as determined by the Freshwater Fish Directive. The values are lower for waters supporting salmonid fishes (salmon, trout) than for cyprinids (bream, carp, tench) because salmonids are less tolerant of poor water quality. (G) is the guideline (recommended) value and (I) is the imperative (mandatory) value.

The pollution of fresh waters has had a great impact on the distribution of freshwater fishes. The term 'pollution' can be difficult to define, but a useful definition is given in Jeffries & Mills (1990) as 'The introduction by humans into the environment of substances or energy that damages life or the physical environment'.

There is a long history of people using the flow of rivers to take away refuse and effluents, and in many places these activities have had an adverse

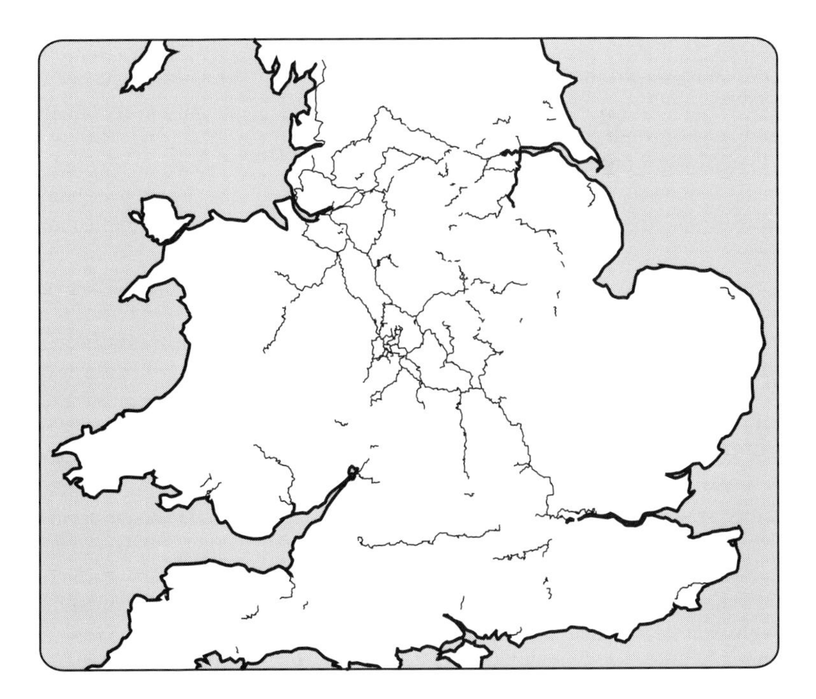

Figure 3
The present-day canal network in England and Wales.

Table 1 Selected water quality parameters from the Freshwater Fish Directive*

Parameter	Salmonid (G)	Salmonid (I)	Cyprinid (G)	Cyprinid (I)
Temperature		21.5°C		28°C
Dissolved oxygen	50%>9mg/l 100%>7mg/l	50%>9mg/l	50%>8mg/l 100%>5mg/l	50%>7
pH		6 to 9		6 to 9
Suspended solids	<25mg/l		<25mg/l	
BOD_5	<3mg/l as O_2			<6mg/l as O_2
Total phosphorus		0.2mg/l as PO_4		0.2mg/l as PO_4
Non-ionized ammonia (NH_3)	<0.005mg/l	<0.025mg/l	<0.005mg/l	<0.025mg/l
Total zinc		<0.3mg/l Zn		<1.0mg/l Zn
Dissolved copper	<0.04mg/l Cu		<0.04mg/l Cu	

Source: Directive 78/569/EEC Directive on the quality of fresh waters needing protection or improvement in order to support fish life ('Freshwater Fish Directive')

* All values are maximum acceptable values, except dissolved oxygen, which are the minimum levels.

impact on fish populations. Since the industrial revolution began in the middle of the eighteenth century, pollution has posed a more widespread and serious threat to fish populations – the effluents from mining, factories and other industrial processes, and the sewage from urban centres has grossly polluted many watercourses (see Chapter 1 – Environmental quality).

The effects of pollution can be extremely complex. Pollutants may act directly on fishes, either acutely (where present in such concentrations as to cause them to die) or chronically (where a pollutant is present at a lower concentration, but sufficient to affect them adversely without immediately killing individuals). An example of acute pollution is a strongly acidic discharge, which would rapidly kill the fishes present in the polluted reach. Chronic toxicity effects can include reduced growth rates, physical deformities, increased susceptibility to diseases and poor reproductive success, all of which are likely to have an adverse impact on fish populations. Substances that disrupt endocrine systems are a good example of chronic pollutants. These mimic the effects of female hormones, feminizing some of the male fishes and can reduce reproductive success in affected populations. Pollution may also impact indirectly, by affecting water quality (for example, by causing oxygen to be removed from the water), by damaging physical habitat (by smothering river gravels etc.), or by affecting other species in the ecosystem (by killing either the invertebrates that fishes feed upon or the plants on which fishes may deposit their eggs).

Although improved legislative controls now strictly regulate the release of effluents from specific discharges, fresh waters are still susceptible to pollution. Accidental or illegal releases usually occur from point sources. There are also diffuse sources of pollution that are difficult to identify and control, such as fertilizer and pesticide run-off from agricultural land, urban drainage and acid rain.

Pollution, whether acute or chronic, diffuse

or from point sources, falls into three main categories according to its constituents: organic, inorganic, and pollution from suspended solids and other physical sources. Thermal, noise and light pollution form separate, distinct categories.

Examples of organic pollutants include dairy waste, domestic sewage, waste from abattoirs and meat processing plants, silage and animal manure. Often, organic pollution is not directly toxic. However, in breaking down the organic matter, bacteria require oxygen that they take from the water. This causes a reduction in the level of dissolved oxygen (an oxygen 'sag'), which in severe cases can remove all the oxygen from a water body, killing fishes and invertebrates in the affected area by suffocation.

Inorganic pollution is caused by chemicals that are directly toxic to freshwater life. These include metals (particularly heavy metals such as cadmium and arsenic), strong acids and alkalis, ammonia, cyanides and many synthetic compounds. They may be directly poisonous to fishes, or act indirectly, for example by killing the invertebrates upon which fishes rely for food. Sources of inorganic pollution are varied; many are waste products from industrial processes.

Physical pollution generally takes the form of silt and other suspended solids. It does not have a chemical action but affects the water environment by physical processes. The nature of the substrate can be changed as gravels are smothered by fine particles, killing water plants, the invertebrates living within and upon the substrate and the eggs of fish such as salmon and trout, which are laid within river gravels. Light penetration can be reduced to such an extent that insufficient light reaches aquatic plants for them to survive, and particulate matter can directly abrade the delicate membranes of the gills of fish. Examples of physical pollution include washings from coal, gravel and other mineral extraction, river engineering works disturbing the channel and re-suspending sediments, vegetable washing and erosion from arable land.

Thermal pollution is produced by the discharge of coolant water from power stations and industrial processes at a higher temperature than the water into which it flows. The effect on the receiving water is to raise the ambient temperature above natural levels. In addition to warming the water, thermal pollution can lead to a decrease in dissolved oxygen levels, whilst also increasing the biological demand of fishes and other aquatic organisms for oxygen. Thermal pollution has a dual impact, benefiting species with a preference for higher temperatures, through enhanced growth and production, whilst excluding species unable to tolerate high temperatures. In some cases, temperatures can be elevated so that alien fish survive in waters that would otherwise be too cold to support them. In the 1960s and '70s, the Leeds–Liverpool Canal in St Helens received heated effluent from a local factory, which allowed cichlids and guppies that had been illegally released into the wild to survive far outside their natural range. The populations persisted for several years, finally dying out only when the heated discharge was stopped.

Noise and light pollution do not affect water quality chemically or physically but may impact on the behaviour of fishes exposed to their effects. The impacts of noise and light, for example from industry or construction activity, can exclude fish from the areas affected because they are disturbed and avoid such areas. This type of pollution can effectively form a barrier to fish movement, prevent fish from making spawning migrations and restrict their distribution. For example, such a behavioural barrier in a river estuary could prevent returning adult salmon and sea trout from entering the river to spawn. The fact that fishes are adversely affected by noise is exploited in the use of acoustic barrier systems, which are employed to exclude fish from water intakes at power stations and other sites. In such situations noise is purposely generated to ensure that fish are kept away from intakes so that they are not removed with the abstracted water.

Eutrophication

Eutrophication refers to the enrichment of water bodies by the input of excess nutrients (mainly compounds of phosphorus). This stimulates an array of changes including increased growth of algae and higher plants, which can adversely affect the diversity of the biological system, the quality of the water, and changes in or, in severe cases, loss of the fish populations present. Sources of nutrient input include:

- Agricultural fertilizers. The run-off of fertilizer from agricultural land following rainfall, or drift of fertilizer directly into the water.

- Sewage inputs. In addition to the problems caused by organic pollution, sewage also contains nutrients that promote eutrophication.
- Detergents. Detergents discharged with waste water contain phosphates.
- Urban run-off. Run-off from urban areas often contains high levels of nutrients. Urban lakes are often eutrophic.

Eutrophication usually takes place in still waters and slow-flowing, lowland rivers. When additional nutrients are added to a system a chain of events is set in motion, typically beginning with the increased growth of algae and aquatic plants responding to higher levels of nutrients available. As planktonic algae proliferate, light penetration through the water column is reduced, preventing sufficient light from reaching rooted aquatic plants, which eventually decline. Blooms of algae cause fluctuations in oxygen levels: they release oxygen in daylight by photosynthesis, when the rate of photosynthesis exceeds the rate of respiration, but remove oxygen at night as they respire. These fluctuations in oxygen levels, and other changes in water chemistry, create stressful conditions for fishes. Oxygen uptake at night can be so severe as to remove virtually all the available oxygen from the water, creating anoxic conditions. When the algal blooms die they decay rapidly due to bacterial action which also removes oxygen from the water and can release toxins injurious to fishes. This affects the structure of the fish community by causing a shift in species composition from pollution sensitive species to those that are pollution tolerant, which may eventually lead to the total loss of all fishes. The Arctic charr of Windermere provide an example of a population affected by eutrophication. Nutrient input from sewage discharges resulted in blooms of algae in the lake, which caused oxygen levels to be depleted, thereby threatening the fishes. Eventually, the eutrophication was reduced through improvements to sewage treatment facilities, and the Arctic charr stock has shown some recovery.

Acidification

The input of additional acid to fresh waters is a process known as acidification. The catchments most susceptible to acidification are those where the natural geology is least able to counteract the effects on water quality of increased input of acids. These conditions are found in areas with a nutrient poor, igneous and metamorphic geology, typical of parts of Scotland, Wales and the south west of England. Catchments with an alkaline geology, such as chalk or limestone, have sufficient quantities of dissolved carbonates and bicarbonates (bases) to react with any additional acidity and to neutralize it. This ability to neutralize acid inputs is termed the 'buffering capacity' of the water.

Catchments that lack adequate buffering capacity become increasingly acidic as more acidifying chemicals enter the system. The majority of acid input to fresh waters in Britain is in the form of 'acid rain', caused by burning fossil fuels (coal, oil, gas, etc.), primarily in power stations and industrial plants. When fossil fuels are burned, they give off various oxides of nitrogen (NO_X) and sulphur dioxide (SO_2), which are released to the atmosphere and ultimately form nitric and sulphuric acid by dissolving in rainwater. These acids are carried by the wind and precipitated as rain or snow, often far from the source of the pollution.

Coniferous plantations can exacerbate this situation because the trees intercept airborne pollution, concentrating the acid and transferring it to ground waters. As the acid percolates through the soil, metal ions are released, notably aluminium, which is itself toxic to fish and other aquatic animals in acidic conditions. Water discharged from abandoned mine workings are also frequently acidic, and in addition can contain high levels of dissolved metals, forming a very polluting effluent.

Fishes are susceptible both to the sharp, periodic increases in acidity caused by acid rainfall, and to the gradual deterioration of water quality over time due to chronic (long term) acidification. Direct effects from acid conditions, and from the increased toxicity of metals in this environment, have served to deplete fish populations severely in affected streams and lakes. The impact of acidification on fishes has been felt in south-west Scotland and in the uplands of Wales and some remedial action has been taken. For example, at Llyn Brianne reservoir on the river Towy, powdered limestone has been added as a temporary solution to help neutralize the acid conditions in the river.

Modifying the river environment

In addition to the natural impacts of drought and flood, human management of water resources modifies the quantity and quality of water and habitat available for fishes.

Water is abstracted from lakes and rivers for domestic supply, agriculture and industry. Over-abstraction from water bodies can cause the same problems as a natural drought: loss of habitat, impeded passage for fishes and deterioration in water quality. Borehole abstraction from ground waters can reduce the height of the water table, making headwaters and streams dry up. Such abstractions are now managed on a catchment basis to take account of the ecological requirements of fishes and other species, which serves to reduce this impact.

The creation of reservoirs in river valleys can have a significant effect on fish populations. In creating a reservoir, river habitat is lost, replaced by still water habitat. Spawning grounds upstream of the reservoir are made inaccessible to returning migratory fish. The flow regime downstream is regulated by the reservoir, often losing the peak flows which remove fine sediments deposited in river gravels. This compromises the success of spawning trout and other species that require clean gravel in which to deposit their eggs. There may not be a sufficient flow of oxygenated water through the gravels because they are clogged with silt, and the eggs deposited within the gravels can be suffocated.

When large volumes of water are released, increased flow velocities may displace fishes and their fry. If water from the cold, poorly oxygenated bottom layers of a reservoir is released, perhaps for hydro-electric power generation, then lack of oxygen or reduced water temperatures may kill fishes, fry and eggs. Following the construction of the Kielder reservoir on the River Tyne, stocks of dace were lost for many kilometres downstream of the reservoir. The release of water much colder than the ambient river temperature, and the displacement of newly hatched fry by high flows are believed to be responsible for this decline. Similar effects have been demonstrated when cold ground waters have been used to support flows in warmer rivers. When a more environmentally sensitive release regime from the reservoir at Kielder was adopted, with lower flow rates and warmer water from the upper layers of the reservoir being released, stocks of dace appeared to recover.

Water transfer schemes, where water from one catchment is transferred to another to augment flow levels, can cause similar effects: higher flow levels and differing water temperature and chemistry can affect fishes in the recipient water. Furthermore, young fishes can be transferred through the pipelines along with the water and colonize the receiving catchment. This is how specimens of spined loach in East Anglia were inadvertently transferred from the Great Ouse to the River Stour.

Reservoirs and much smaller structures, such as weirs, mill dams, bridge footings and fords, also affect fish populations by restricting access to areas of the catchment upstream. These obstructions most obviously affect migratory species, such as salmon and sea trout, and fish passes are frequently built to allow them access to their spawning grounds in the upper reaches of catchments. Nevertheless, on many historic structures, the need for fish passes was not considered at the time of construction, and migratory fishes were denied access to large areas of catchments in which they were formerly present. Coarse fishes and eels, whose commercial value is much lower than that of salmon and trout and which are less able to jump than salmonids, have not generally received the benefit of fish passes designed to facilitate their passage. Increasingly, migration has been shown to play an important part in the reproductive cycles of coarse fish species and barriers preventing them from migrating to spawning areas inevitably have an adverse impact on populations.

In managing flood risks, many watercourses have in the past been modified to increase their capacity to store and carry floodwater. This typically involves techniques such as deepening and widening the channel by dredging, removal of fallen trees and other obstructions, straightening the channel to remove meanders and channel realignment to reduce the roughness of the banks. These changes to the channel reduce the diversity of habitats available (including backwaters and areas away from the main current) and food availability, and destroy spawning substrates, and therefore cause fish populations to be severely depleted or lost. In recent years, increased understanding of the effects of this type of river engineering has led to a number of river restoration projects, including the Little Ouse in Norfolk, the Dearne in South Yorkshire, the Skerne in County Durham and the Ash in Surrey. Meanders and other natural features have been restored, banks and channels re-profiled and vegetation re-introduced to recreate the natural habitat these rivers once offered, allowing fishes to recolonize the restored areas.

Modifying and maintaining watercourses so that they are suitable for navigation employs some of the methods used in flood management, with the same effects on fish populations. Channels may be dredged and widened to ensure sufficient depth and width of water for boat traffic, or be straightened to allow longer craft to negotiate bends, and in-stream and riparian vegetation may be removed to improve access. The creation of weirs and locks, which intermittently impound watercourses, can also modify the species of fishes present by providing habitat for species that favour slower, deeper water.

The direct effects of navigation also impact upon fish populations, mainly through the effect of wash from boat traffic. The advancing wake and backwash generated by boats can erode bank sides, increase turbidity and uproot aquatic plants. Aquatic plants are also sometimes removed to maintain a navigable channel for boats. In severe cases, almost all submerged vegetation may be removed from navigable reaches. Loss of aquatic plants reduces the spawning substrate available for species such as roach and bream, which deposit their eggs on water plants, and removes habitat used for cover and feeding by both adult and juvenile fishes.

Fisheries Management

In addition to the incidental effects on fish distribution of the activities discussed above, the active management of fisheries has significantly affected the distribution of fishes in Britain.

One of the most popular management activities is the stocking of fishes – there are approximately 10 million fishes stocked to over 3000 different locations each year in England and Wales alone. There is nothing new in this practice: the common carp was introduced into Britain as a source of food in the latter part of the fifteenth century. Carp, along with pike and perch and other species were kept in monastery and manor ponds to supply the table. Carp stocking continued and was later taken up by anglers, who value them for their size and sporting qualities, and the common carp now has a very widespread distribution, entirely as a result of human influence.

Stocking, particularly with species alien to Britain, can have adverse effects on the community of native species, and is therefore strictly controlled by law. Notable examples of stockings that have caused adverse effects include the zander, first introduced to two lakes in the Woburn Estate, Bedfordshire, in 1878. Zander from Woburn were introduced to the Great Ouse Relief Channel in the 1960s and from there have spread rapidly through East Anglia and, through illegal introductions, have now reached major river catchments such as the Thames, Severn and Trent. Because the zander preys mainly on other fishes, introductions have had a deleterious effect on stocks of the indigenous species where they were introduced. Similarly, the introduction of ruffe into Loch Lomond in the 1980s (probably through anglers releasing surplus livebait) is believed to have had a detrimental effect on the native whitefish population of the loch, since part of the ruffe's diet includes both the eggs and young of whitefish. Ruffe introduced to Bassenthwaite Lake and Derwent Water in the Lake District are also considered to pose a threat to vendace populations there.

Fish stocking does not always cause damage to native stocks, and has been used in many cases to restore fish populations to rivers recovering from the effects of long-term pollution. The River Tyne, once so grossly polluted that the native salmon stock was all but destroyed, has, by significant improvements to water quality in the estuary, coupled with an extensive salmon stocking programme, been transformed into the premier salmon river in England and Wales.

Further reading:

General – Giles, 1994; Greenhalgh, 1999; Maitland & Campbell, 1992; Phillips & Rix, 1985; Wheeler, 1983.

Origins – Lamb, 1995; Wheeler, 1977.

Natural distribution – Huet, 1949, 1959; Varley, 1967.

Human influences – Cowx, 2001; Hynes, 1970; Jeffries & Mills, 1990; Mason, 1981.

3 The Database and Atlas of Freshwater Fishes project

Work on the Database and Atlas of Freshwater Fishes (DAFF) project was initiated in 1996 and continued until March 2002. It has resulted in a database of records of the occurrence of freshwater fishes in Britain (including the Isle of Man and the Channel Islands) and Northern Ireland, that forms the basis of this Atlas. For reasons explained later in this chapter, the Atlas itself does not include Northern Ireland. Here we describe the DAFF project, why it was undertaken and the reasons for its scope and the methods used. We provide a brief review of the sources of data and the ways in which datasets were obtained. We also describe how a summary of the DAFF database is being made accessible via the Internet.

Why the DAFF project was initiated

Freshwater fishes are the only vertebrate group that, at the end of the twentieth century, did not have an up-to-date collated source of information on the distribution of species occurring in Britain. This information is essential to many organizations, such as those concerned with nature conservation and environmental protection, as well as anglers and general naturalists. Several species of freshwater fishes are currently listed for protection under national or international legislation or conventions: we need to know where and when they occur to be able to safeguard them. The commoner species are also important. Information is needed about where individual species occur to be able to study them, to preserve stocks or, in the case of some introduced species, to control their spread. Information about individual species can also be combined with other data, for example about land use, other organisms, water quality or the management of rivers, canals, reservoirs and lakes, to interpret the interaction of fishes with their environment.

As early as 1988, the Biological Records Centre (BRC) identified a need to fill this particular gap in basic knowledge of our biodiversity, but it was not until 1996 that resources could be found to begin the work. The DAFF project was set up by a partnership of organizations to produce a source of collated information on the distribution of all freshwater and estuarine fish species (including anadromous species) throughout Britain and Northern Ireland.

Other national databases and atlases

During the 1990s, atlases were published covering the distribution of birds, mammals, amphibians and reptiles in Britain, or in Britain and Ireland. Similar atlases of mosses and liverworts, aquatic plants and several invertebrate groups, such as butterflies, dragonflies, grasshoppers and molluscs, were also published in the same period. All these atlases are summaries of more detailed information in computerized databases that, in many cases, are managed by BRC.

The only complete set of detailed distribution maps for fishes was that published in 1972 by the Freshwater Biological Association in Peter Maitland's *Key to British freshwater fishes*. These maps cover Britain and Ireland showing distribution at the scale of 10km squares. Unlike most modern atlases, Maitland's maps were not supported by a computerized database and the mechanical process by which the original maps were prepared at BRC in 1971 had become obsolete by 1972.

Although Maitland compiled an outstanding summary of the information available at the time, it is inevitable that the maps were based on incomplete coverage, with some areas lacking records of any species. In their New Naturalist volume, *Freshwater fishes*, Maitland and Campbell included small scale 'range' maps for most species in Britain and Ireland, and made an unequivocal plea for a 'sophisticated mapping scheme' and a centralized source of distribution data to support both policy and practice in nature conservation. There are several recent examples of range maps for freshwater fishes in Europe, often at a small scale, that include species in Britain and Ireland.

Feasibility study 1996

The project began in 1996 with a feasibility study on the preparation of a national dataset and an *Atlas of freshwater fishes*. This study was undertaken by Paul Harding at BRC and Ian Winfield of the (then) Institute of Freshwater Ecology (IFE). The study was funded jointly by the Centre for Ecology and Hydrology (CEH) (then known as the Institute of Terrestrial Ecology (ITE)) and the Joint Nature Conservation Committee (JNCC). The chief aim was to define the priorities, opportunities and problems associated with the compilation of a national dataset and the publication of an atlas. Other staff at IFE were involved in this study, and a representative of the former National Rivers Authority (now the Environment Agency) and several fisheries experts were also consulted.

The study concentrated on the potential for collating and abstracting relevant data from existing datasets and making greater use of these, rather than attempting to set up new national-scale data-gathering initiatives. The study concluded that it would be practicable to compile a national summary database and prepare an atlas. Success would depend on the willingness of the potential sources of data to contribute at least summary records. It was also necessary to obtain sufficient resources to carry out the work over a period of three to five years. The majority of potential sources of data were organizations rather than individuals. Given the coverage achieved by Maitland in 1972, and the amount of survey work done since then, it was not unreasonable to expect that coverage would be at least as good as that achieved for most 'popular' groups, other than birds or butterflies.

Pilot study 1997

A pilot study was undertaken by BRC in 1997, based on the principle that the database and atlas project should not initiate new surveys of freshwater fishes. It would draw on existing, dispersed datasets to collate a minimum set of data to answer a limited range of questions about where and when freshwater fishes have been recorded in Britain and Northern Ireland. The pilot study began to establish formal agreements and technical routes for access to existing datasets and to assess what types of data it would be useful and possible to compile in a unified, national summary database. A questionnaire survey was sent to the holders of existing datasets to identify those for inclusion in the database and to find out what information was available.

Minimum information requirements

The pilot study established that certain information would be essential for the production of an atlas of species distribution. The minimum requirement for data 'fields' included:

- Name of the organism (e.g. species, subspecies and hybrids);
- Date of the record;
- National Grid reference;
- Some other geographical location information to enable validation of the grid reference;
- Name of the person (or organization) responsible for identification (to enable validation of the record).

Other data fields were also collated to allow a wider range of scientific uses of the completed database.

The DAFF project 1998–2002

The DAFF project was funded by CEH, JNCC and the Environment Agency and data collation was carried out over the period April 1998 to March 2002. The DAFF database is not intended to provide a compendium of information on fishes because the FishBase database already provides access to information on the biology and ecology of all freshwater and marine species. FishBase is now accessible in CD ROM form and on the Internet (see Appendix 4). The DAFF database collated the minimum fields required for distribution mapping plus those fields that the questionnaire highlighted as being possible and useful to collate. It did not collate the full range and variety of data held by its contributors. However, a summary index to relevant data (a form of metadata) was compiled as a by-product of the project, as each potential source of data was examined and the data collated.

Sources of data

Datasets were acquired from a variety of sources, such as local record centres, museums, water companies and the personal records of individual specialists. By far the greatest volume of data for England and Wales was obtained from records held by the Environment Agency and its predecessors, including all the accessible data from surveys. The history of the Agency meant that at the time the DAFF project was running, each regional fisheries department had its own methodology and there was no unified method of collating fisheries data between regions. DAFF was able to obtain the information from these diverse sources for its own purposes, but subsequently the Agency has developed an integrated system for managing fisheries data.

Data from Scotland have been obtained chiefly from the Scottish Fisheries Co-ordination Centre, FRS Freshwater Laboratory and Peter Maitland. Information on the distribution and abundance of coarse fishes in Scotland has lagged behind that available for England and Wales. To go some way to redressing this, the Scottish Executive recently commissioned survey work, mainly by local Fisheries Trust biologists, to provide better information on the current position, but this information was not available.

A small quantity of recent and historical data was obtained from the Channel Islands, mainly from Guernsey Museum and Art Gallery and the Environmental Services Unit, Jersey.

The format and content of the datasets that were collated was almost as disparate as the number of datasets supplied. The challenge for the DAFF project was to extract the relevant data from each contributing source and to ensure that as far as possible there were no gaps in coverage either geographically, or in terms of species data, and that the temporal coverage was as good as possible. A summary of the data used to produce Maitland's 1972 maps has been included in the database to enable comparisons to be made with more recent data.

Ireland

Unlike many comparable atlases of species distribution, the geographical coverage of this atlas is limited to Britain (with the Isle of Man and the Channel Islands), but does not include either Northern Ireland or the Republic of Ireland. However, a review of the status of vertebrates in Ireland, compiled by A. Whilde and published in 1993, covered freshwater fishes and drew on available sources of information on threatened species throughout the whole island.

Although the DAFF project collated many recent and historical data from Northern Ireland, the overall coverage remains patchy and, as a consequence, maps including Northern Ireland could be misleading. Nevertheless, Northern Ireland data have been included in the DAFF database (see below). The Centre for Environmental Data and Recording at the Ulster Museum in Belfast is now collating records of freshwater fishes in Northern Ireland.

There are, of course, many potential sources of data in the Republic of Ireland, including the Central Fisheries Board, the seven Regional Fisheries Boards and numerous technical and scientific publications, but the DAFF project team ascertained at an early stage that collated, electronic datasets were not available to provide adequate coverage of the occurrence of freshwater fishes in the Republic.

Access to the DAFF database

A summary of the DAFF database, and the associated metadata, are accessible via the Internet Gateway of the National Biodiversity Network (NBN) (see Appendix 4). This enables anyone with access to the Internet to search for a summary of the records of any species and includes data from Northern Ireland that are not included in this atlas.

The NBN Gateway makes the DAFF database available for use by the general public in a variety of ways, including mapping at 10km square resolution, checking scientific and common names of species and linking to other web sites with information about freshwater fishes. Most of the records are also available with more details than just 10km square summaries, using the interactive mapping facility of the NBN Gateway. For example, the distribution of species can be mapped at the more detailed spatial scale of the original records (see Figure 4, p. 34). However, it should be noted that we are prevented from giving access to some records at a more detailed level because, in a few cases, the

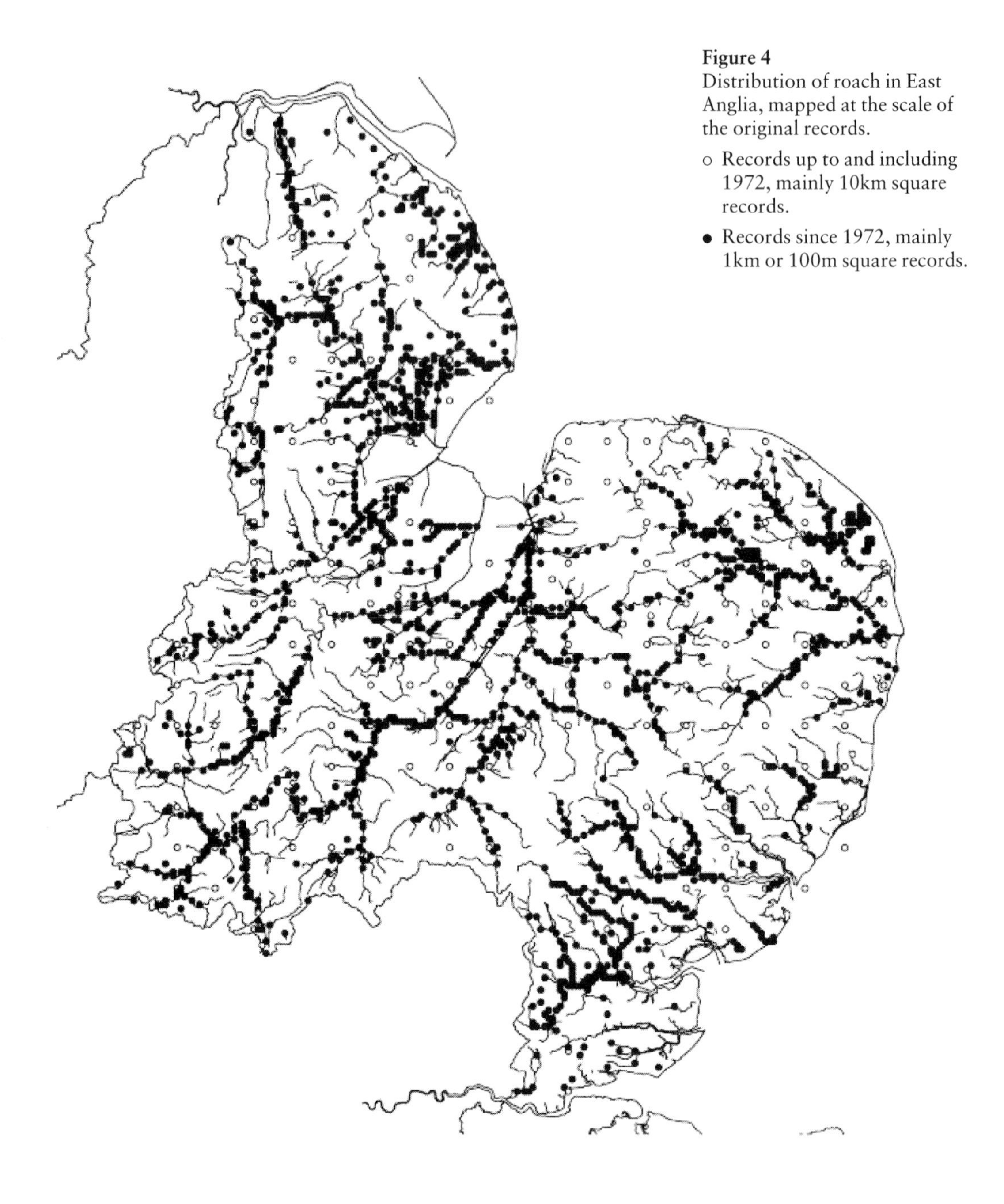

Figure 4
Distribution of roach in East Anglia, mapped at the scale of the original records.

- o Records up to and including 1972, mainly 10km square records.
- • Records since 1972, mainly 1km or 100m square records.

suppliers of those records have requested that their detailed data should not be made publicly available. Anyone wishing to gain access to the complete data for any species or area is advised to contact the suppliers of records whose details are contained in the relevant metadata on the NBN Gateway.

Further reading:

Greenhalgh, 1999; Maitland, 1972, 1990, 2000; Maitland & Campbell, 1992; Miller & Loates, 1997; Whilde, 1993.

Plate 1 (*see p. 11*):
Fly-fishing for Atlantic salmon.
River Don, Aberdeenshire.

Plate 2 (*see p. 11*):
Angling competition.
River Torne, Humberhead
Levels.

Plate 3 (*see p. 12*): (*Left, top*) Electrofishing to monitor the fishery in a wide, deep river using a boom-boat with several electrodes. River Tees near Stockton-on-Tees, Co. Durham.

Plate 4 (*see p. 12*): (*Left, below*) Atlantic salmon being released by angler. River Dee, Aberdeenshire.

Plate 5 (*see p. 12*): (*Right*) Commercial fishing for Atlantic salmon and sea trout in a tidal estuary using the traditional net and coble method. River Tweed, near Berwick-upon-Tweed, Northumberland.

Plate 6 (*see p. 15*): (*Below*) Commercial fishing for Atlantic salmon using a seine net. Photographed in the late 1980s, this fishery no longer exists. River Itchen at Wood Mill, Hampshire.

Plate 7 (*see p. 17*): Angler beside waterfalls in a mountain stream. Water of Feugh, near Banchory, Deeside, Aberdeenshire.

Plate 8 (*see p. 17*): Electrofishing in a small stream for fisheries research using a single anode, with the generator on the bank in the foreground. Pill River, north Somerset.

Plate 9 (*see p. 17*): Electrofishing in a shallow stream for fisheries research using a power source in a backpack, Bindon Millstream, beside River Frome, Dorset.

Plate 10 (*see p. 21*): Atlantic salmon leaping: Buchanty Spout, River Almond, Perthshire.

Plate 11 (*see p. 23*): Trout zone in the fast-flowing, well-oxygenated headwaters of the River Severn. Afon Gam, tributary of Afon Banwy, near Llangadfan, Powys.

Plate 12 (*see p. 23*): Barbel zone in a slow-flowing lowland river, populated by barbel, chub and other species. River Great Ouse, Clifton Reynes, Buckinghamshire.

Plate 13 (*see p. 24*): An oligotrophic lake, low in the nutrients required for aquatic plant and algal growth and with few species of fish. Wast Water SSSI, Lake District National Park, Cumbria.

Plate 14 (*see p. 25*): A small canal connected to the network of canals and rivers throughout midland England, from the Mersey to the Thames and the Severn to the Trent. It is designated as an SSSI and SAC. Cannock Extension Canal, Staffordshire.

Plate 15: (*see p. 142*)
A classic English chalk stream famed for its brown trout, rich invertebrate fauna and many species of aquatic plants.
River Test SSSI, Chilbolton Common, Hampshire.

Plate 16: (*see p. 142*)
A small lowland river with outstandingly rich aquatic vegetation and macro-invertebrate fauna and with many species of fish.
Moors River SSSI (a tributary of the Dorset Stour) at Troublefield Copse, near Bournemouth, Dorset.

4 Species accounts and distribution maps

Introduction

P. S. Maitland's distribution maps 1972

The distribution of freshwater fishes in Britain (and Ireland) was poorly known until Peter Maitland set up a national recording scheme for freshwater fishes in 1966, which he described in 1969 and 1970 in a series of publications. Maitland worked in collaboration with the Biological Records Centre (BRC) and the scheme was formalized through an Advisory Panel on Mapping the Distribution of Freshwater Fish set up in April 1967 and chaired by David Le Cren. The initial results were summarized as maps of each species published in Maitland's *Key to British freshwater fishes* (1972). Although these were intended as interim maps, they have remained the only detailed national distribution maps for all the British species until this publication. Maitland continued to receive records of freshwater fishes after the maps were produced for publication, but the scheme and its Advisory Panel did not continue formally after about 1970. The expected demand from anglers for Maitland's *Key* and its maps never materialized and this edition remained in print and on sale until 2001. It is a tribute to Maitland's contribution to knowledge of the distribution of our freshwater fishes that his original data, suitably updated, have formed the basis of information on the status of freshwater fishes in Britain throughout the last 30 years. A revised and updated edition of this work has been published in 2004.

The maps compiled for Maitland's 1972 *Key* used a mechanical process based on 40-column punched cards. This process became obsolete soon after it was published and, for reasons that have never been satisfactorily explained, the mechanical mapping cards that covered the fishes were not converted to digital data when similar data sets were converted by BRC in the mid 1970s. Most of the original record cards were retained by Maitland, but other cards, compiled mainly by BRC and comprising some 2000 individual records, have been retained at BRC. The data set for the 1972 maps has been recreated at BRC as part of the DAFF project.

DAFF database

It was not until 1996 that resources were found to begin work on producing an up-to-date and comprehensive collated information source on species distribution. The history of the DAFF project and its scope are described in Chapter 3. The project collated data from a wide variety of sources and includes results from many fisheries surveys, as well as data collected for other purposes.

Geographical limitations of data

Geographical coverage is comprehensive but there are certain areas where either no records have been submitted, or where there are relatively few. Figure 5 (p. 36) illustrates the coverage of the data. Dot symbols indicate that there is at least one record for that 10km square; blank squares show that no information has been submitted. The vast majority of records in the database are from flowing water (rivers and streams) as Environment Agency and Scottish Fisheries Co-ordination Centre (SFCC) surveys of fisheries target these more frequently. Still water records, from natural lakes, reservoirs and other man-made fisheries, are in some areas sadly lacking, which has led to the described distribution of still water species being under-representative.

Temporal limitations of data

The earliest records in the database are from 1637, for eel and chub in the Sheffield district. There are no further records in the database until the middle of the eighteenth century, for which a scatter of records has been obtained from published reports and museum archive material. Records up to the end of the nineteenth century are sparse, although a block of records for the 1790s was acquired from the Old Statistical Account of Scotland. Several of the standard eighteenth- and nineteenth-century books on

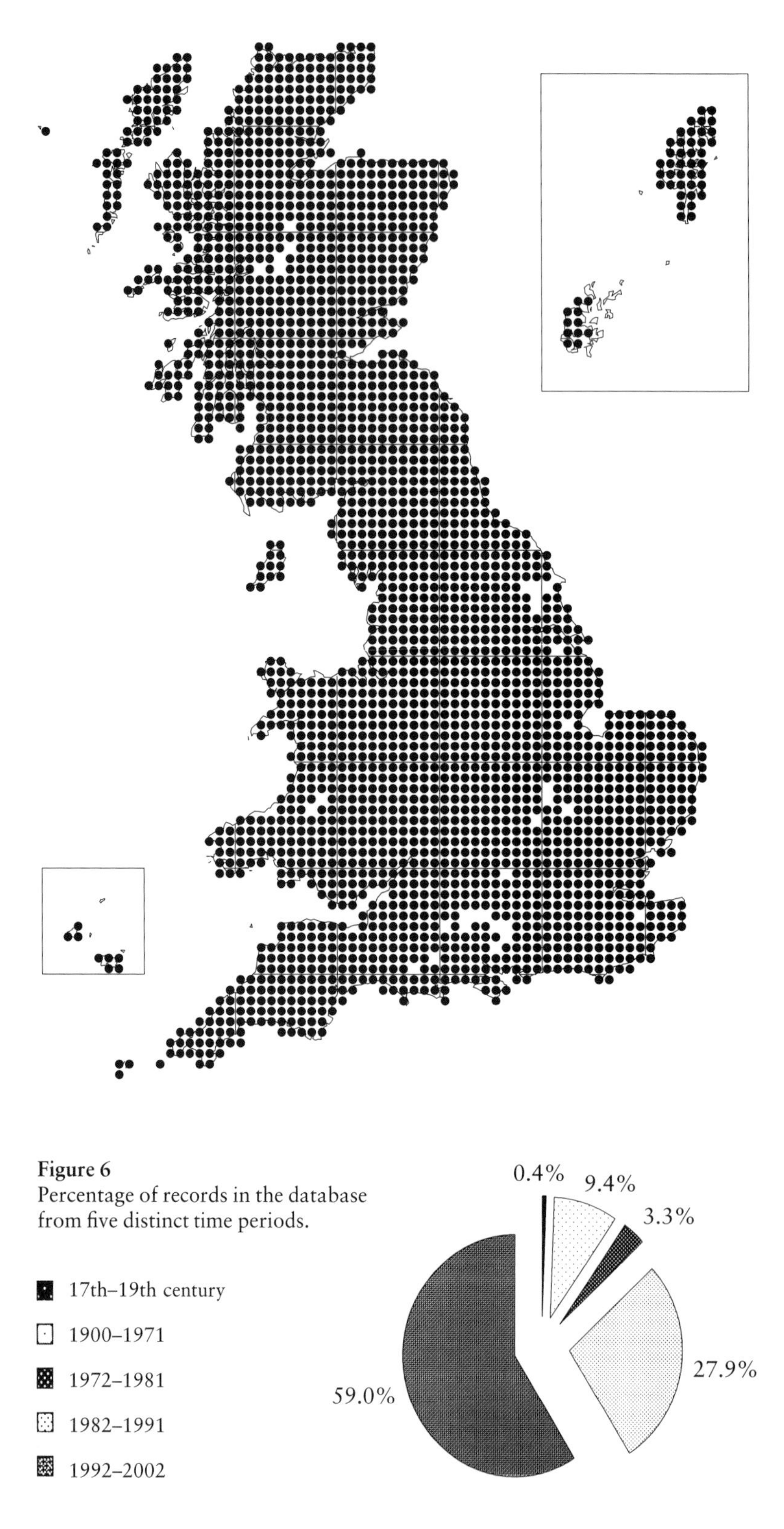

Figure 5
Coverage of data in database mapped at 10km square; dot symbols indicate at least one record per 10km square.

Figure 6
Percentage of records in the database from five distinct time periods.

British fishes (see Appendix 1) include the localities where fish were known to be present, but these data have not been extracted as part of the project. In addition, there is almost certainly a large quantity of historical data in old fishing and natural history books and in museum collections and archives that has not been collated into the DAFF database. These historical records usually lack precise localities and are difficult to access, so they have not been actively sought for the project. Data collection started in earnest only when Peter Maitland began his recording scheme in 1966. After the publication of his *Key*, the amount of recording declined for several years. Environment Agency data from the mid-1980s for England and Wales have been made available to the DAFF project, with increasing amounts of data becoming accessible with more regular use of computerized systems from about 1990 onwards. Figure 6 shows the percentage of records in the database from five distinct time periods.

Taxonomic scope of this book

This chapter summarizes the results of the DAFF project and includes species accounts and 10km square distribution maps for all native as well as many introduced species. All fishes commonly encountered in British fresh waters have been included. Estuarine species able to penetrate the upper reaches of an estuary or into tidal rivers have also been included, but those restricted to coastal waters have been excluded.

Some species, such as lampreys and grey mullets, are difficult to separate to species in the field, especially as juveniles. In many cases, survey work has identified these only to genus, particularly where the species concerned are not of economic or sporting importance. For these reasons, combined accounts and maps have been produced for lampreys (sea, river and brook lamprey), and for grey mullets (thick-lipped, thin-lipped and golden grey mullet). There has been some disagreement among taxonomists whether brown trout and sea trout should be given sub-specific status. Survey work, particularly for juvenile salmonids, does not distinguish between brown trout and sea trout and so the map accompanying the species account is for all *Salmo trutta* records, although where sea trout have been identified that information is stored in the database.

Many coarse fish species are capable of interbreeding, as they share similar breeding times and habitat preferences resulting in individuals of different species spawning together in close proximity. Roach × common bream hybrids are the most common, and together with other hybrids they are sometimes recorded in fisheries' surveys. These fish, which combine the physical characteristics of both parent species, can be difficult to identify, even for experienced fisheries' workers. Hybrid fish may be misidentified as either of their parent species; consequently their recorded distribution is believed to reflect their true distribution less accurately than that of non-hybrids. For this reason, although records of the distribution of hybrids have been recorded in the database, their distribution has not been mapped here.

The diversity of fish in Britain is increasing, particularly with the introduction of non-native species for sport, some of which have become established in the wild. Several of these recent non-native introductions have been included, where they are present in numbers sufficient to be noted in routine surveys or where they have been notified through licensing agreements or still-water fishery owners' questionnaire returns. There are many other introduced species of which specimens may be encountered in Britain and for which data may be held in the database. Maps have not been included for these species as the data are likely to be incomplete and would be misleading. In addition, many of these non-native introductions are present only temporarily in a particular water, often as isolated specimens; in other cases the continued presence of populations is entirely dependent on repeated stocking.

Houting (*Coregonus oxyrinchus*)

This species is included in the check list for completeness, but no map or species account is included here, and there are no records of houting in the DAFF database. In the area of the southern North Sea, houting was historically associated with large and medium-sized rivers from at least the Netherlands to southern Denmark, and it occurred occasionally in the rivers of south-east England as a vagrant. However by the 1980s, as a result of pollution, river re-engineering and the construction of weirs, the only remaining population in the North Sea area was that of the Vidaa River in western Denmark, close to the German border.

Houting is the only anadromous whitefish of the North Sea area. It is morphologically and physiologically distinct from other whitefish of this area, of which two species, vendace (*C. albula*) and powan (*C. lavaretus*) occur in Britain. Houting differ from other whitefish in having an elongated upper jaw and being adapted to high salinities. Genetic work, however, has shown that houting is very closely related to *C. lavaretus*, and has evolved relatively recently, perhaps within the past 13,000 years. It is not yet clear whether houting should be recognized as a separate species distinct from *C. lavaretus*, but the special morphology and physiological adaptations should justify its high priority for conservation.

Check list of freshwater fish species in Britain

A check list of fish species found in Britain, together with their status (native or otherwise) is given in Table 2. The taxonomic sequence follows that of Nelson (1994). Nomenclature follows that used in Wheeler (1992) with minor exceptions. Where there is a departure from Wheeler's list, this is noted in the table. Commonly used vernacular names are also included: for a more comprehensive list of vernacular names see Maitland & Campbell (1992), and for international vernacular names see Maitland (2000).

Table 2: Check list and vernacular names

Scientific name	Vernacular name	Status
Petromyzontidae		
Lampetra fluviatilis (L., 1758)	River lamprey or lampern	Native
Lampetra planeri (Bloch, 1784)	Brook lamprey	Native
Petromyzon marinus L., 1758	Sea lamprey	Native
Acipenseridae		
Acipenser sturio L., 1758	Common sturgeon	Native
Anguillidae		
Anguilla anguilla (L., 1758)	European eel	Native
Clupeidae		
Alosa alosa (L., 1758)	Allis shad	Native
Alosa fallax (Lacépède, 1803)	Twaite shad	Native
Cyprinidae		
Abramis bjoerkna (L., 1758)[1]	Silver bream or white bream	Native
Abramis brama (L., 1758)	Common bream	Native
Alburnus alburnus (L., 1758)	Bleak	Native
Barbus barbus (L.,1758)	Barbel	Native
Carassius auratus (L., 1758)	Goldfish	Introduced
Carassius carassius (L., 1758)	Crucian carp	Native?
Ctenopharyngodon idella (Valenciennes, 1844)	Chinese grass carp, grass carp or white amur	Introduced
Cyprinus carpio L., 1758	Common carp (incl. leather carp, mirror carp, ghost carp, king carp and koi carp)	Introduced

[1] Kottelat (1997) in FishBase. Wheeler (1992) uses the synonym *Blicca bjoerkna* (L., 1758).

Scientific name	Vernacular name	Status
Gobio gobio (L., 1758)	Gudgeon	Native
Hypophthalmichthys molitrix (Valenciennes, 1844)[2]	Silver carp	Introduced
Leucaspius delineatus (Heckel, 1843)[3]	Sunbleak, motherless minnow or belica	Introduced
Leuciscus cephalus (L., 1758)	Chub	Native
Leuciscus idus (L., 1758)	Orfe or ide (incl. golden orfe, blue orfe)	Introduced
Leuciscus leuciscus (L., 1758)	Dace	Native
Phoxinus phoxinus (L., 1758)	Minnow	Native
Pseudorasbora parva (Temminck & Schlegel, 1846)[4]	Topmouth gudgeon or false harlequin	Introduced
Rhodeus sericeus (Pallas, 1776)	Bitterling	Introduced
Rutilus rutilus (L., 1758)	Roach	Native
Scardinius erythrophthalmus (L.. 1758)	Rudd	Native
Tinca tinca (L., 1758)	Tench	Native
Cobitidae		
Cobitis taenia L., 1758	Spined loach	Native
Balitoridae		
Barbatula barbatula (L., 1758)[5]	Stone loach	Native
Ictaluridae		
Ameiurus melas (Rafinesque, 1820)	Black bullhead	Introduced
Siluridae		
Silurus glanis L., 1758	Danube catfish or wels	Introduced
Esocidae		
Esox lucius L., 1758	Pike	Native
Osmeridae		
Osmerus eperlanus (L., 1758)	Smelt or sparling	Native
Salmonidae		
Sub-family Coregoninae		
Coregonus albula (L., 1758)	Vendace	Native
Coregonus autumnalis (Pallas, 1776)[6]	Pollan	Ireland only
Coregonus lavaretus (L., 1758)	Whitefish, powan, schelly or gwyniad	Native
Coregonus oxyrinchus (L., 1758)[7]	Houting	Vagrant, believed extinct

[2] Added by Kottelat (1997), not in Wheeler (1992).

[3] From Skelton (1993) in FishBase, not in Wheeler (1992).

[4] From Bogutskaya & Naseka (1996) in FishBase, not in Wheeler (1992).

[5] Kottelat (1997); Nelson (1994). Wheeler (1992) puts this in family Cobitidae.

[6] Occurs in Northern Ireland and the Republic of Ireland, but not in Britain.

[7] Kottelat (1997) in FishBase. Wheeler (1992) does not give it species status but includes with *C. lavaretus*. A rare vagrant in the 19th Century, now believed extinct in Britain. Hansen *et al.* (1999) describe the genetic relationships of *Coregonus* species.

Scientific name	Vernacular name	Status
Sub-family Salmoninae		
Oncorhynchus gorbuscha (Walbaum, 1792)	Pink salmon	Introduced
Oncorhynchus mykiss (Walbaum, 1792)	Rainbow trout	Introduced
Salmo salar L., 1758	Atlantic salmon	Native
Salmo trutta L., 1758	Brown or sea trout	Native
Salmo trutta subsp. *fario*[8]	Brown trout	Native
Salmo trutta subsp. *trutta*[8]	Sea trout	Native
Salvelinus alpinus (L.,1758)	Arctic charr	Native
Salvelinus fontinalis (Mitchill, 1814)	Brook charr or American brook trout	Introduced
Sub-family Thymallinae		
Thymallus thymallus (L., 1758)	European grayling	Native
Gadidae		
Lota lota (L., 1758)	Burbot	Native
Gasterosteidae		
Gasterosteus aculeatus L., 1758	Three-spined stickleback	Native
Pungitius pungitius (L., 1758)	Nine-spined or ten-spined stickleback	Native
Cottidae		
Cottus gobio L., 1758	Bullhead	Native
Moronidae		
Dicentrarchus labrax (L., 1758)[9]	Bass, sea bass	Native
Centrarchidae		
Ambloplites rupestris (Rafinesque, 1817)	Rock bass	Introduced
Lepomis gibbosus (L., 1758)	Pumpkinseed or pumpkinseed sunfish	Introduced
Micropterus salmoides (Lacépède, 1802)	Largemouth bass	Introduced
Percidae		
Gymnocephalus cernuus (L., 1758)	Ruffe or pope	Native
Perca fluviatilis L., 1758	Perch	Native
Sander lucioperca (L., 1758)[10]	Pikeperch or zander	Introduced
Mugilidae		
Chelon labrosus (Risso, 1826)	Thick-lipped grey mullet	Native
Liza aurata (Risso, 1810)	Golden grey mullet	Native
Liza ramada (Risso, 1826)	Thin-lipped grey mullet	Native
Cichlidae		
Tilapia zillii (Gervais, 1848)	Tilapia or redbelly tilapia	Introduced
Pleuronectidae		
Platichthys flesus (L., 1758)[11]	Flounder	Native

[8] Not all authorities accept the sub-specific status, but it is felt useful to distinguish the information where it is available.

[9] Nelson (1994); Smith (1990) in FishBase. Wheeler (1992) retains this in family Percichthyidae.

[10] Kottelat (1997) in FishBase. Wheeler (1992) uses the synonym *Stizostedion lucioperca* (L., 1758).

[11] Cooper & Chapleau (1998) in FishBase. Wheeler (1992) prefers the synonym *Pleuronectes flesus* (L., 1758).

Interpreting the distribution maps

The distribution map accompanying the species accounts shows the distribution of that species or species group at 10km square resolution based on records in the DAFF database. The two areas shown in boxes on the maps are the Channel Islands (lower left) and the Orkney and Shetland Islands (upper right). The Channel Islands are mapped on the international UTM grid, all other areas are mapped on the National Grid. A record is the reported occurrence of the species at a site on a particular date. All records falling within a particular 10km square at any time within a specified time span are plotted as a single point in the middle of that 10km grid square. As the central point of some coastal squares falls in the sea this can give the appearance of some freshwater species being reported offshore! Records of the location of the species are usually held at a finer spatial resolution in the database, usually to 100m and much of this more detailed information will be accessible via the Internet using the National Biodiversity Network (NBN) Gateway (see Appendix 4). The species accounts in this book summarize key facts about the individual species. For further information, there are many useful references given in the bibliography to which people can turn, and larger public libraries may carry some of these even when they are no longer in print. Staff at aquaria and museums are often well-informed and will assist with enquiries; in particular, museum staff can often be helpful in using keys and other reference material to identify fish specimens correctly. However, progressively, people are turning to the Internet for further information. Of the general fish information sources available through the Internet, FishBase (see Appendix 4) is particularly useful for its key facts, life history information, photographs and extensive references on all fish species.

Table 3 below provides the number of records in the DAFF database for each species (as at June 2002). For the maps, the data have been 'time-sliced' showing records from two distinct time periods: records in the database up to and including 1971, and records from 1972 to 2002 inclusive. The circle symbol ○ shows 10km squares where there are records only from the earlier time period, while dot symbols ● indicate records only from 1972 onwards. The square symbol ■ distinguishes 10km squares where there are records for both time periods, indicating some continuity of recording. The grey-shaded squares underlying the other symbols represent the distribution of that species as published in Maitland's 1972 *Key*. Some squares show only the grey shading, as no validated record for that species in that square has been made available to the project. The apparent changes in distribution shown by the different symbols must be interpreted with caution, as the amount of recording taking place has increased dramatically since 1972, with about ten per cent of the records in the database falling within the earlier time period (see Figure 6, p. 36). Thus apparent increases in range or density are likely to reflect the increase in survey intensity, although some real changes in range may be distinguished for some species. For example, ruffe have extended their range into Scotland in recent times and grayling and rudd have been introduced into some Scottish waters where they did not previously occur. Figure 7 (p. 43) shows the number of different species per 10km square which gives an indication of species richness. Four classes are identified using equal interval classification.

Table 3 Number of records per species in the DAFF database

Name	Number of records	Name	Number of records
Abramis bjoerkna	3407	*Alosa fallax*	189
Abramis brama	10,218	*Ambloplites rupestris*	1
Acipenser sturio	57	*Ameiurus melas*	1
Alburnus alburnus	3,021	*Anguilla anguilla*	27,060
Alosa alosa	133	*Barbatula barbatula*	13,388

Name	Number of records
Barbus barbus	2,014
Carassius auratus	264
Carassius carassius	1,472
Chelon labrosus	88
Cobitis taenia	736
Coregonus albula	62
Coregonus lavaretus	127
Cottus gobio	14,332
Ctenopharyngodon idella	227
Cyprinus carpio	3,717
Dicentrarchus labrax	189
Esox lucius	15,766
Gasterosteus aculeatus	10,722
Gobio gobio	13,790
Gymnocephalus cernuus	5,298
Lampetra fluviatilis	569
Lampetra planeri	1,639
Lepomis gibbosus	22
Leucaspius delineatus	59
Leuciscus cephalus	10,300
Leuciscus idus	131
Leuciscus leuciscus	13,979
Liza aurata	7
Liza ramada	8
Lota lota	45
Micropterus salmoides	4
Oncorhynchus gorbuscha	4
Oncorhynchus mykiss	2,401
Osmerus eperlanus	427
Perca fluviatilis	17,429
Petromyzon marinus	383
Phoxinus phoxinus	11,234
Platichthys flesus	2,579
Pomatoschistus microps	178
Pseudorasbora parva	6
Pungitius pungitius	953
Rhodeus sericeus	200
Rutilus rutilus	22,252
Salmo salar	25,621
Salmo trutta	22,881
Salvelinus alpinus	613
Salvelinus fontinalis	67
Sander lucioperca	1,392
Scardinius erythrophthalmus	4,335
Silurus glanis	168
Sprattus sprattus	101
Thymallus thymallus	2,990
Tilapia zillii	1
Tinca tinca	5,943
Subspecies	
Salmo trutta subsp. *fario*	18,572
Salmo trutta subsp. *trutta*	2,002
Hybrids	
Abramis brama × *Abramis bjoerkna*	32
Abramis brama × *Rutilus rutilus*	4,374
Abramis brama × *Scardinius erythrophthalmus*	5
Alburnus alburnus × *Leuciscus cephalus*	3
Alburnus alburnus × *Leuciscus leuciscus*	2
Abramis bjoerkna × *Scardinius erythrophthalmus*	1
Carassius auratus × *Carassius carassius*	1
Carassius auratus × *Cyprinus carpio*	2
Carassius carassius × *Cyprinus carpio*	1
Leuciscus cephalus × *Leuciscus leuciscus*	3
Leuciscus cephalus × *Rutilus rutilus*	55
Leuciscus leuciscus × *Rutilus rutilus*	3
Rutilus rutilus × *Abramis bjoerkna*	64
Rutilus rutilus × *Scardinius erythrophthalmus*	342
Salmo trutta × *Salmo salar*	5
Genus only	
Alosa	190
Lampetra	2,672
Metriaclima	2
Family only	
Mugilidae	15
Surveys with no fish recorded	599

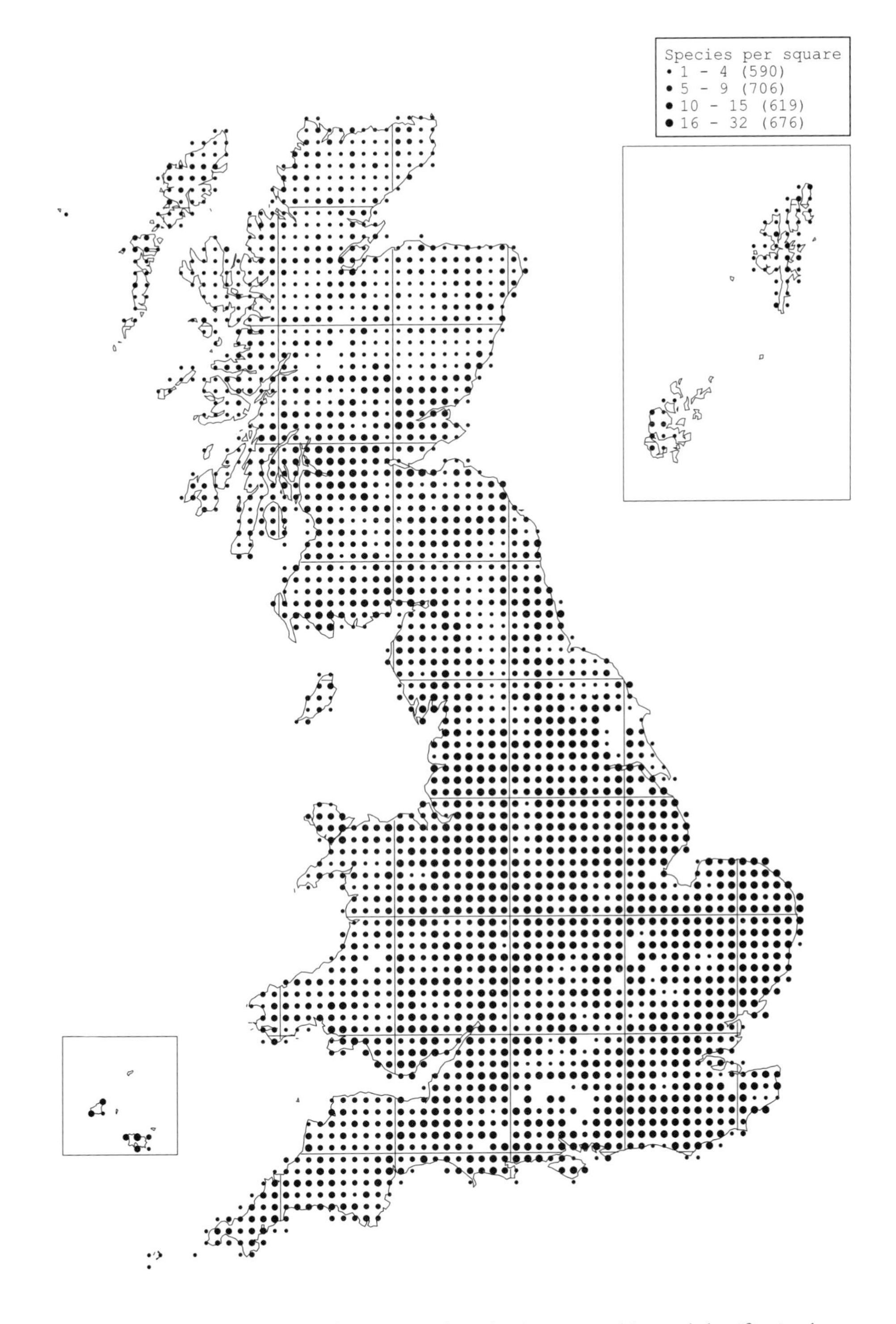

Figure 7 Number of taxa per 10km square, plotted using an equal interval classification by dividing all 10km squares into four roughly equal groups, based on the number of species recorded in each square.

Lampreys
(family *Petromyzontidae*)

Three species are recorded in Britain of which two, the sea lamprey *Petromyzon marinus* and the river lamprey *Lampetra fluviatilis*, are anadromous, whilst the brook lamprey *Lampetra planeri* remains in fresh water for its entire life-cycle.

Description

- Primitive, jawless, eel-like fishes, without paired fins or scales.
- Skeleton composed of cartilage, unlike the teleost (bony) fishes.
- Seven gill pores behind each eye and single nostril on top of head.
- Distinctive round, suckered mouth, containing teeth in adults.
- Juvenile forms (ammocoetes) are blind, worm-like, grey-brown in colour. Gill pores evident, the mouth without sucker.
- Ammocoetes metamorphose into pre-adults, with fully developed eyes and suckers. Silvery grey to olive on dorsal surfaces with almost white underbelly.
- Sea lamprey

This is the largest of the anadromous, parasitic lampreys. On entering fresh water, they are steely grey in appearance, with dark, often black mottling. As they approach spawning the mottling fades and they attain a less distinct brown coloration.

- River lamprey

On entering fresh water, river lampreys' coloration can vary from green-grey to steely grey on their dorsal surfaces, but they lack the distinctive mottling of the sea lamprey. The ventral surfaces are almost white. As they approach spawning, they change to an olive-brown colour, slightly lighter on the ventral surfaces. The river lamprey is unusual in that it is known to exist in land-locked populations, as for example at Loch Lomond.

- Brook lamprey

After metamorphosis, the adults are olive brown on the dorsal surfaces, sometimes silvery along the flanks, with a light underbelly.

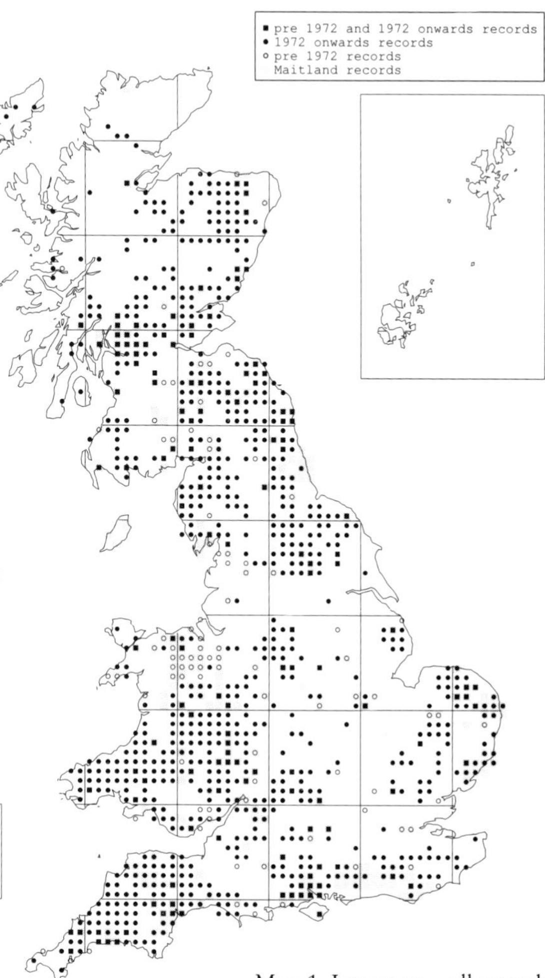

Map 1 Lampreys – all records

Size

Sea lamprey: adults reach 100cm.

River lamprey: adults normally 30 to 40cm, but specimens of nearly 50cm have been recorded. The Loch Lomond race is smaller than the truly anadromous forms, attaining a maximum length of only 25cm.

Brook lamprey: adults normally 15cm, but up to 19cm in some populations.

Biology and behaviour

All the lampreys require clean, flowing water with a loose gravel and pebble substrate to spawn successfully. In Britain, river and brook lampreys spawn in spring through to early summer (March to June), whilst sea lampreys may spawn into

July. All the lampreys spawn in nests, which are depressions in gravel, with a distinctive mound at the downstream end. They create a nest initially by using their suckers to remove stones which, with the aid of the current, they deposit downstream of the spawning site. The nest is completed using vigorous body movements, which causes particulates to be loosened and washed downstream. Lampreys often do this whilst attached by the sucker to a suitably sized pebble. Brook and river lampreys may spawn in groups of up to 50 individuals, but river lampreys have also been observed spawning in much smaller groups of two females escorted by up to six males. The sea lamprey is generally believed to spawn monogamously, with the male arriving at the spawning site first to initiate spawning activity. It has recently been shown that reproductively mature male sea lampreys release a bile acid that acts as a powerful sex pheromone, active over long distances, which attracts females to spawning sites. Male sea lampreys are aggressive and will defend their spawning site. All lampreys die after spawning.

Once the eggs hatch, the larvae drift downstream to suitable silt beds where they live in burrows feeding on microscopic organisms. River lampreys have an average pre-metamorphic life of four and a half years, whilst sea lamprey ammocoetes

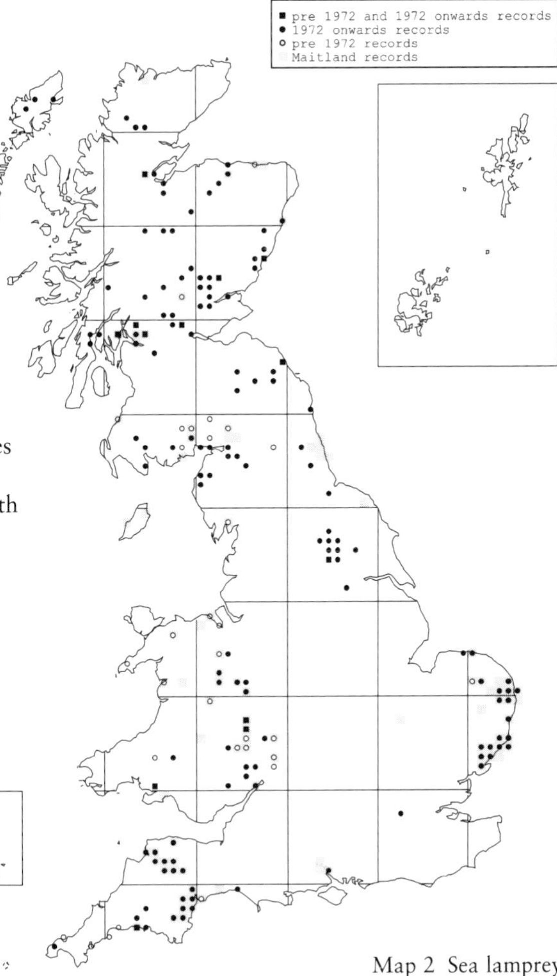

Map 2 Sea lamprey

Sea lamprey

generally develop for six to eight years. Brook lamprey ammocoetes develop for about six years, before metamorphosing to the adult form during the autumn. Once they are adult, brook lampreys do not feed, but live for approximately six months.

The anadromous species migrate to the sea during the autumn and winter where they feed by attaching their powerful suckers to host fishes, rasping through the skin to gorge on flesh and blood. River and sea lampreys feed on a variety of hosts including members of the cod and herring families, whilst the Loch Lomond population of river lamprey are known to feed on whitefish (powan). Sea lampreys have been observed attached to whales and other cetaceans and are also reported to have attached themselves to boat hulls in the North Sea.

After the marine phase (which lasts about eighteen months for river lampreys and up to three years for sea lampreys) adults migrate back to fresh water to complete the life-cycle. Unlike salmon, which migrate back to their natal streams, the homing behaviour of anadromous lampreys is poorly understood. Tagging of emigrating sea lampreys has indicated that adults do not return to natal waters but are instead attracted into watercourses in which sea lamprey ammocoetes are present. It is unclear how once-extinct populations of migratory lampreys are replaced and what prompts them to return to watercourses without lampreys.

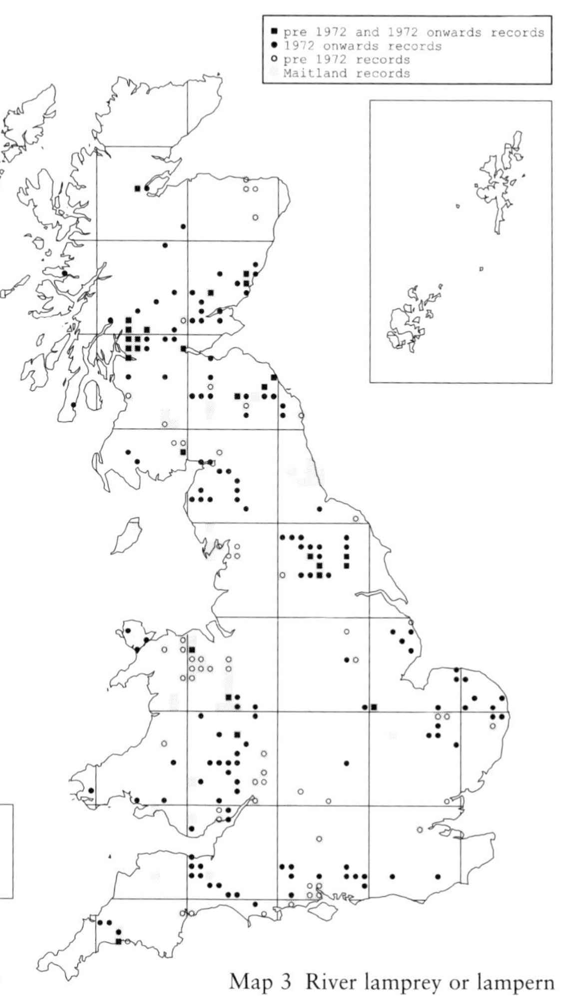

Map 3 River lamprey or lampern

Habitat

The ammocoetes are relatively sedentary, living in burrows in the silty river bed, but displacement by floods often gives rise to mixing of populations. They are fairly tolerant of a range of temperatures, but 29°C is recorded as an upper,

River lamprey or lampern

lethal limit for brook lampreys. Brook lampreys are typically found in small, shallow streams, which are often the headwaters of major river systems, whilst ammocoetes and pre-adults of river and sea lampreys have been recorded in relatively deep lowland rivers. Runs of sea lampreys are particularly associated with medium to large rivers and they can penetrate well inland to spawn where rivers are not obstructed.

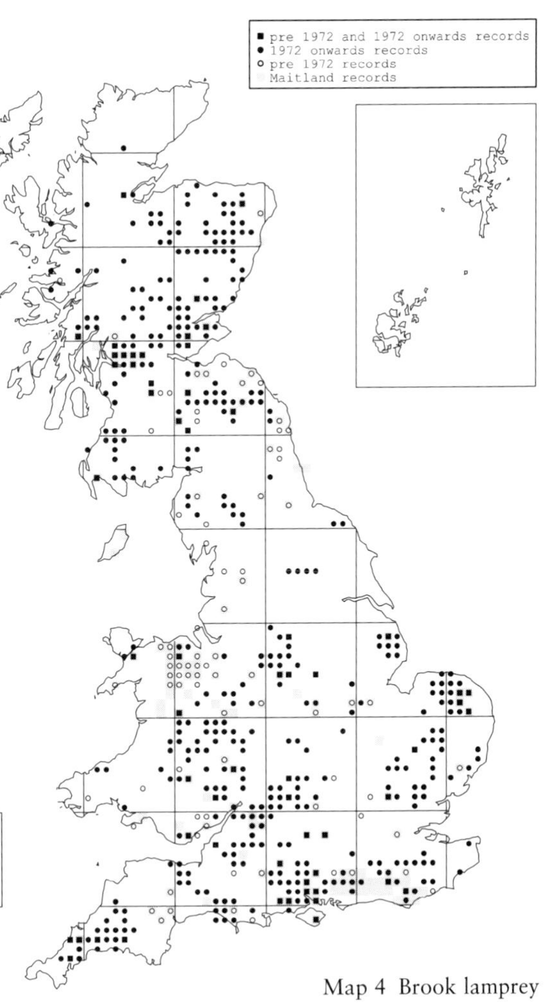

Map 4 Brook lamprey

Distribution in Britain

Lampreys, although fairly widespread throughout Britain, are secretive and rarely observed but, as awareness increases, more observations are being recorded. Improvement of the water quality of tidal rivers during the last 30 years has assisted in the recovery of anadromous populations.

The sea lamprey is found mainly throughout the coastal river systems of East Anglia, the South-West and south and east Scotland. The rivers Wye and Severn are particularly good strongholds, together with several rivers in Wales. The Yorkshire Ouse and Derwent systems have seen notable increases in sea lamprey numbers in the 1990s, and in 2000 they were observed in the river Thames, believed to be the first records in modern times. These colonizations are thought to be due to improvements in water quality. The largest change in the recorded distribution since 1972 has been the increase of records in Scotland, primarily as a result of extensive survey work commissioned by Scottish Natural Heritage. In England, more accounts have been received from their previous strongholds of Wales, East Anglia and the South-West. Sea lampreys appear to be absent from much of north-west England and the Midlands, with only sparse records from the southern counties.

Brook lamprey

Due to the difficulties of differentiating between ammocoetes of the river and brook lampreys, many recorders have listed their observations merely as *Lampetra* sp. These are included in the combined distribution map (Map 1) and can be compared with those of the individual species. The river lamprey is widespread throughout Britain, with considerable overlaps in range with the sea lamprey, but it is also recorded in Anglesey, Lincolnshire and the southern counties. The brook lamprey has the most widespread distribution of all the lampreys, extending from small populations on a few of the Western Isles to strongholds in north-east Scotland, although it is rare north of the Great Glen. It occurs more extensively in the English Midlands and southern counties than either the river or sea lamprey.

World distribution

The sea lamprey is widely distributed on both sides of the North Atlantic and occasionally recorded as far north as Greenland; it is sporadic in northern Europe, with some records from northern Norway. It occurs in the coastal waters of southern Norway and Sweden, but infrequently off Finland. Due to pollution and migration barriers, it is becoming rarer in central Europe, but large populations are still recorded in western Europe, particularly in French rivers draining to the Atlantic. It is common in the rivers of the Iberian Peninsula, but much less abundant across the Mediterranean. It is found in coastal rivers of Italy and the Balkan coast of the Adriatic and along the coast of North Africa. The construction of shipping seaways into the Great Lakes of North America gave the sea lamprey access where it proliferated causing extensive damage to the indigenous fish populations with the collapse of commercial fisheries in some of the lakes.

The river lamprey is more restricted than the sea lamprey, ranging from southern Norway, through central Europe across to the French Atlantic coast, where the distribution appears to cease along the Iberian coast. There are small populations across the Mediterranean and the Italian Adriatic, and land-locked populations in Finland and Russia.

The range of the brook lamprey overlaps that of the river lamprey, but the former generally occurs further inland in the upper catchments, often beyond the migration range of the river lamprey.

Status

- All three species are protected under the Habitats and Species Directive and the Bern Convention and they are also listed under the UK Biodiversity Action Plan (see Appendix 2).
- All three species are threatened throughout their range by habitat degradation and the anadromous forms are particularly affected by industrial pollution and physical barriers to upstream migration such as locks, weirs and dams.

Hybrids and related species

- The sea lamprey has no known natural hybrid forms.
- River and brook lamprey hybrids have been produced by artificial fertilization. Electrophoretic analysis of both species indicates a very high degree of similarity. This gives rise to the argument that the river and brook lamprey are paired species with a close phylogenetic lineage including anadromous and non-anadromous forms.
- The juvenile forms of both the river and brook lamprey are indistinguishable and can live in sympatric populations.
- The family Petromyzontidae includes about 35 species worldwide. The monospecific genus *Petromyzon* is most closely related to the genus *Caspiomyzon*, and *Lampetra* shares similar taxonomy to the genus *Lethenteron*.

Lampreys and Man

Whilst historically little culinary importance is attached to the brook lamprey, Romans, Vikings and mediaeval Britons all held sea and river lampreys in high regard as an epicurean delight. In Chapter VIII of Book XIV of *Magiae naturalis*, 1589, Giambattista della Porta gives a recipe: '*...let him be turned, always moistening the fillets with strewing on the Decoction of Origanum. When part of it is roasted ... it will be a gallant meat...*'.

One of the most frequently quoted 'tales' about lampreys is that Henry I (and possibly also King John) '*died from a surfeit of lampreys*'. Although Henry was already in poor health at the time, he died in Anjou in France, so English lampreys are unlikely to have been the cause of his death! Nevertheless, lampreys were a noble food and the lamprey fishers of the Severn in Gloucester presented baskets of lampreys to the sovereign annually. Several other commercial lamprey fisheries around Britain are well documented, but most suffered a decline in lamprey catches as stocks were affected by pollution and barriers to migration. There also appears to have been a decline in their popularity as food in Britain. In the nineteenth century much of the declared catch from the Yorkshire Ouse and Thames fisheries was sent to the Dutch market to be used by anglers for bait in the North Sea cod fishery rather than for domestic use. The Yorkshire Ouse Fishery regularly exported two to four tonnes of river lamprey annually, but it closed in the early twentieth century as a result of pollution.

River lampreys are eaten today in France and Spain, where commercial fisheries still exist, and in Scandinavia and the Netherlands, where smoked lamprey is considered a delicacy. A small commercial fishery for river lampreys has started again on the Yorkshire Ouse, taking up to two tonnes annually which, paradoxically, are supplied to anglers in Britain as bait for pike.

Anglers do not consider the lampreys as sport fishes and they are not targeted in rod-and-line recreational fisheries. However, sea lampreys have been taken recently by anglers, whilst fishing with bait for other species, including on worm and, bizarrely, on luncheon meat!

Further reading:

Beamish, 1980; Bergstedt & Seelye, 1995; Bianco & Muciaccia, 1982; Frear & Shannon, 1994; Hardisty & Potter, 1971; Holčík, 1986; Hubbs & Potter, 1971; Li *et al.*, 2002; Lucas & Baras, 2001; Potter & Beamish, 1975; Porta, 1589; Van Utrecht, 1959.

Author: Paul Frear, Environment Agency

Common sturgeon
Acipenser sturio

Description

- Instantly recognizable 'prehistoric' appearance, with five distinctive rows of bony plates (scutes) running length of body, but without scales.
- Dorsal, anal and spined ventral fins set back towards asymmetric tail, where backbone forms part of upper lobe of caudal fin.
- Extended snout with two pairs of ventral barbels and underslung, extendable mouth.
- Back dark grey/black fading to lighter underside, often with green or yellow.
- Juveniles have all morphological features of adults, but bony plates lighter.

Size

Can grow up to 3m long and weigh over 200kg.

Biology and behaviour

Maturing sturgeon return to their natal rivers in the spring and search out suitable spawning grounds, typically deep gravel-bed pools, in the lower reaches. The female, according to her size, can deposit between 400,000 and 2,000,000 sticky eggs, which attach to gravel, rocks or submerged wood. Hatching takes around five days depending on water temperature.

Juveniles feed on benthic invertebrates, and downstream migration normally occurs after two years. They continue to grow at sea, feeding on molluscs, crustaceans and worms, and there is evidence that sturgeon consume small fishes. Maturity is reached at an age of about ten years when around one metre in length. The adults stop feeding prior to their migration back to fresh water and as a result mortality after spawning is common. However, repeat spawning is possible and large specimens, more than 40 years old, have been recorded.

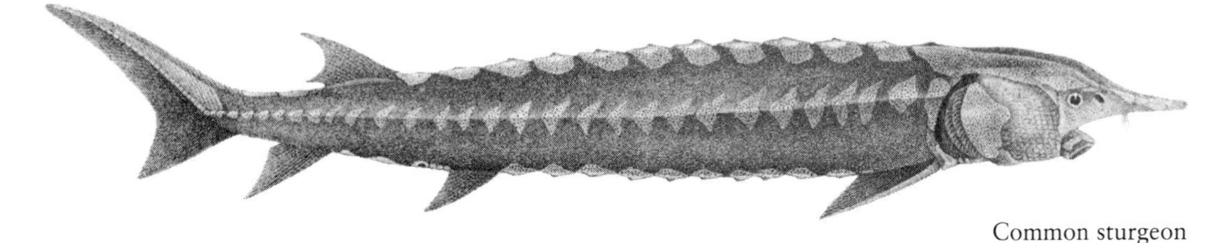

Common sturgeon

Habitat

In their natal river, young sturgeon are principally bottom feeders preferring a loose bed substrate such as sand or mud. Little is known about the adults' marine preferences: some individuals reside in shallow waters of their 'home' estuary whilst others have been caught hundreds of kilometres away by commercial fishermen trawling at depths to 100m.

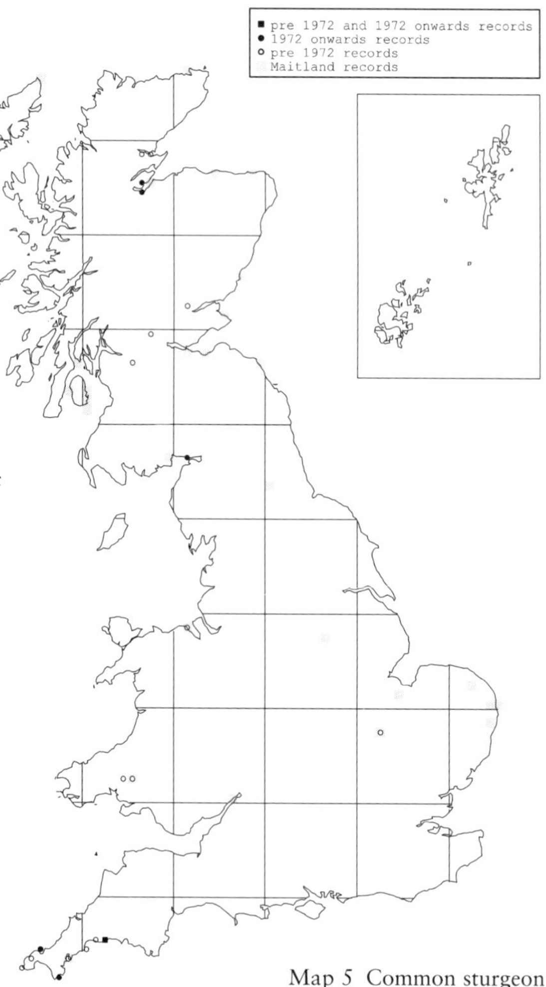

Map 5 Common sturgeon

Distribution in Britain

The common sturgeon is the only sturgeon species ever to have been caught in British waters. Commercial netsmen in coastal waters were, and are, most likely to encounter the species as part of their bycatch. The species does not spawn in our rivers, so any sturgeon found in fresh water in Britain are considered 'vagrants'.

The data collected for the DAFF database indicate that only thirteen individuals were observed or captured in Britain during the twentieth century. Seven records were from Scotland, and three each from England and Wales; most were recorded in river estuaries. Inland records are much scarcer; in the 1930s anglers landed two from rivers in south Wales, one of which weighed more than 200kg. The capture from Hemingford Grey in Cambridgeshire, in 1860, was some 65km from the sea. It was not the first, or last, sturgeon to find its way into the river Great Ouse. Local newspapers of the time reported that for six weeks in the summer of 1924 a sturgeon was seen in this stretch of river until it was netted by local people. Residents said that it was the fifth 'monster' fish to have been captured in this way. There are also scattered records of sturgeon scutes from archaeological sites, such as at the mediaeval Eynsham Abbey, close to the river Thames near Oxford.

World distribution

The former range of the common sturgeon was along the entire European coastline including the Baltic, Mediterranean and Black Seas. Although the species has an anadromous life history, a purely freshwater population was found in Ladozhskoye Ozero (Lake Ladoga) in Russia.

English name does not appear in some atlases.

Breeding used to occur in several European rivers, such as the Rhine, Elbe and Danube; however, these spawning grounds have been lost, and the last spawning population is thought to be in the river Gironde in France. Successful reproduction was recorded there in 1988 and possibly in 1995.

The total number of surviving common sturgeon is impossible to estimate. Adult fishes are still caught occasionally in the Atlantic off the coast of France and in the North and Irish Seas. There were 179 incidental captures between 1980 and 1994, although 125 were recaptures.

Status

- Classified as *critically endangered* by IUCN, protected under CITES, Habitats and Species Directive, Bern Convention, Wildlife and Countryside Act, and Conservation Regulations, and listed under the UK Biodiversity Action Plan (see Appendix 2).
- Susceptible to river pollution and over-fishing. Deterioration of the habitat at spawning grounds and barriers to migration have also contributed to the decline of a species that is particularly vulnerable to such changes due to the length of time before it first reproduces.
- All species of the genus *Acipenser* are listed on the Prohibition of Keeping or Release of Live Fish Orders (see Appendix 3) as species for which release to the wild is not permitted without a licence.

Hybrids and related species

- The common or European Atlantic sturgeon is one of about 27 species found in the Northern Hemisphere and is thought to be capable of interbreeding with several other sturgeon species. It is unclear how many distinct species there are; several sub-species have also been described.
- Six other sturgeon species are found in eastern European waters. Other species of sturgeon inhabit both coasts of North America and the rivers of northern Asia that drain to the Pacific.
- Sterlet, *Acipenser ruthenus* Linnaeus, a native of eastern Europe is regularly imported into the UK for sale through the aquarium trade.

Sturgeon and Man

The common sturgeon now has little commercial value, compared to other species, such as the Beluga, Osetra and Sevruga sturgeons of the Black and Caspian Seas, which are prized for the preparation of caviar (the eggs preserved with salt) from the ovaries of mature females. The North American species of sturgeon are favoured as sport fishes.

All sturgeon caught in British waters should be offered to the reigning monarch, a tradition said to have started because Edward II was particularly fond of the roe. The flesh can also be eaten, and has been described as 'a compound of veal and eel, with the flavour of lobster'.

Author: Roger Handford, Environment Agency

European eel *Anguilla anguilla*

Description

- Long, thin, cylindrical body with dorsal and anal fins merged together, no pelvic fins, very small scales embedded in an extremely slimy skin.
- Unpigmented juveniles (glass eels) are found in estuaries in early spring. On entry to fresh water, glass eels become progressively pigmented (elvers). In river eels the dorsal surface is generally dark, brownish-green with ventral surface much lighter in colour, sometimes light yellow (yellow eels).
- Maturing eels, as they migrate seawards, become very dark on back and silvery or white underneath (silver eels).

European eel

- Eels can be distinguished from lampreys by their jawed mouth, paired nostrils and paired pectoral fins.

Size

Males normally 30–40cm long, females up to 55cm; weight up to 5kg.

Biology and behaviour

European eels are thought to spawn at depth in the Sargasso Sea in early spring, after which the adults die. The Sargasso Sea is part of the North Atlantic Basin, south-west of Bermuda and east of the island chain from the Bahamas to Puerto Rico. It is assumed that the eggs drift with the prevailing Gulf Stream, hatching and developing into the leaf-like, transparent leptocephalus larvae. These young eels are so different in appearance to their parents that for 40 years they were classified as an entirely different species, and the truth was discovered only in the last years of the nineteenth century.

The juvenile eels are carried in the warm currents of the Gulf Stream to the shores of Europe and during the journey the leptocephalus larvae metamorphose into glass eels. There are differing opinions on how long this journey takes, but probably at least a year. The glass eels enter rivers in spring, using the tidal flow, and actively swim upstream provided the water temperature is above 6°C.

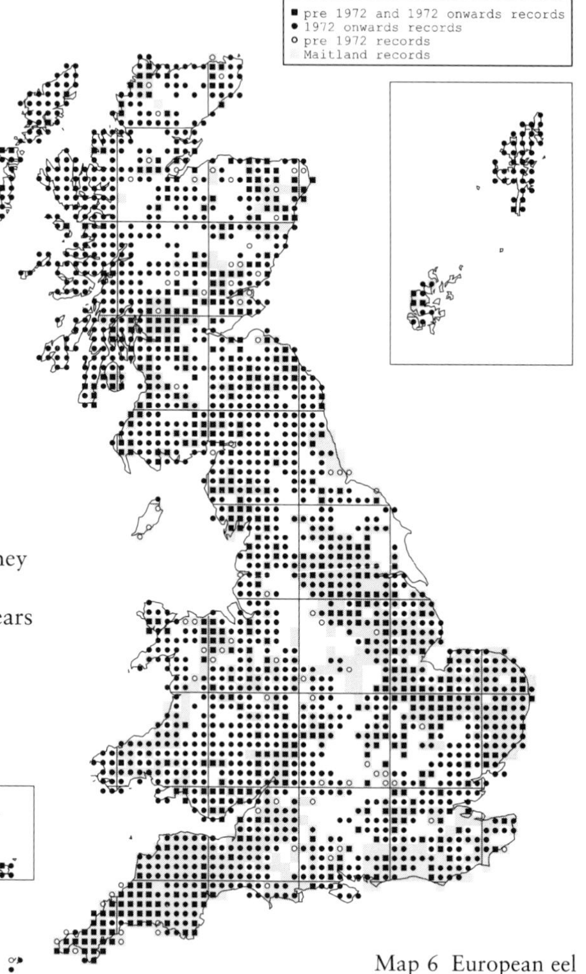

Map 6 European eel

Once in fresh water their behaviour changes quickly as pigmentation proceeds. At this stage, the river eel tends to hide during the day and feed at night, mainly on snails, insect larvae and small, bottom-living fishes. Movement up-river from the tidal zone to the headwaters is slow and very dependent on temperature; males tend to dominate the lower reaches and the larger females the middle and upper reaches. Eels remain in fresh water for several years before beginning their return migration and specimens over 40 years old have been recorded.

One of the most notable aspects of eel migration is their ability, alone amongst British fishes, to migrate overland. Eels have frequently been observed moving over wet terrain, usually at night. They achieve this by closing their gill

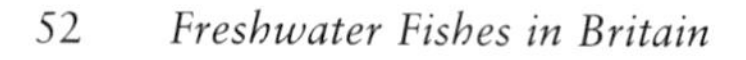

covers, thereby trapping a reserve of water in their gill chambers. This ability probably accounts for their widespread distribution.

The return migration to sea takes place on dark, stormy nights in autumn; males migrate at a younger age and size (aged six to twelve years, and less than 40cm in length) whilst females are older and larger (aged eight to fifteen or more, and longer than 40cm). Virtually nothing is known about the adult marine phase but the eels make their way back to their spawning grounds in the Sargasso Sea to begin the cycle once more.

Habitat

Eels can be found in almost all fresh waters: rivers, lakes, ponds, canals, ditches, estuaries and coastal areas all provide suitable habitats. In fresh water, they seem to prefer lowland lakes and rivers where there is plenty of cover, the water is still or slow-moving, and the bottom is muddy, although they can also be found in upland rivers and mountain streams. They are tolerant of wide ranges of temperature, pH and dissolved oxygen.

Distribution in Britain

Eels are almost ubiquitous although in some river systems access has been prevented by large obstructions, both man-made and natural, or as a result of pollution. Large estuarine and coastal populations also exist.

World distribution

There are at least 17 species of 'freshwater eel' worldwide. The closest relative to the European eel is the American eel *Anguilla rostrata* which is also migratory from the Atlantic. The European eel occurs from the Faroe Islands and Iceland, through northern, western and southern Europe including the Mediterranean and Black Seas, to the Atlantic coast of North Africa.

Status

- Current information indicates a major decline in numbers of glass eels during the last two decades throughout Europe, including Britain. The European Commission recognizes the problems with eel replenishment, stocks and fisheries and may act to protect the stock to safeguard fisheries for the future.
- Long term datasets at a few key sites in Britain indicate no impact yet on adult eel stocks, but there is concern over trends.

Hybrids and related species

- It has always been assumed that European eels are part of a single breeding population. More recently, there has been some evidence of genetic differentiation into northern (i.e. Icelandic), western European and Moroccan groups.
- No other freshwater eel species or hybrids occur in Europe.

Eels and Man

All the life stages are exploited. The principal glass eel and elver fisheries are in tidal reaches of the river Severn and other rivers flowing into the Bristol Channel. Yellow eels are exploited using traps and nets in many areas, although East Anglia is the main centre for this activity. Silver eels are also taken by means of traps, fyke nets, trawls, and fixed fishing weirs. Elver catches in Britain are believed to be about ten tonnes and those of yellow and silver eels to be a few hundred tonnes. The largest European elver fisheries are in France, but there are also some in Spain, Portugal and Britain. There is an important elver fishery on the River Bann at Toomebridge, Northern Ireland. Many of these elvers are restocked elsewhere in the Bann catchment, principally Lough Neagh, for harvesting as yellow or silver eels. The principal market for elvers is the Far East, for aquaculture. On the Continent, most yellow eels are caught in mainland Europe and most silver eels in the Baltic. Small quantities of frozen eels are imported into Britain each year, mainly from China and New Zealand, for direct consumption.

Further reading:

Sinha & Jones, 1975; Tesch, 1977.

Authors: Alan Churchward & Jonathan Shelley, Environment Agency

Shads – *Alosa* species

Two species of *Alosa* occur in Britain, the allis shad *A. alosa* and the twaite shad *A. fallax*. They are very similar in general appearance and biology. Both species spawn in rivers but spend most of their life at sea. They occur all around the coasts of Britain, but records of breeding populations in fresh waters rely on identification of adults to species level, or of eggs and fry, which regrettably are often identified only to genus. Twaite shad breed in several rivers in Britain but there are no confirmed recent records of allis shad breeding in our rivers.

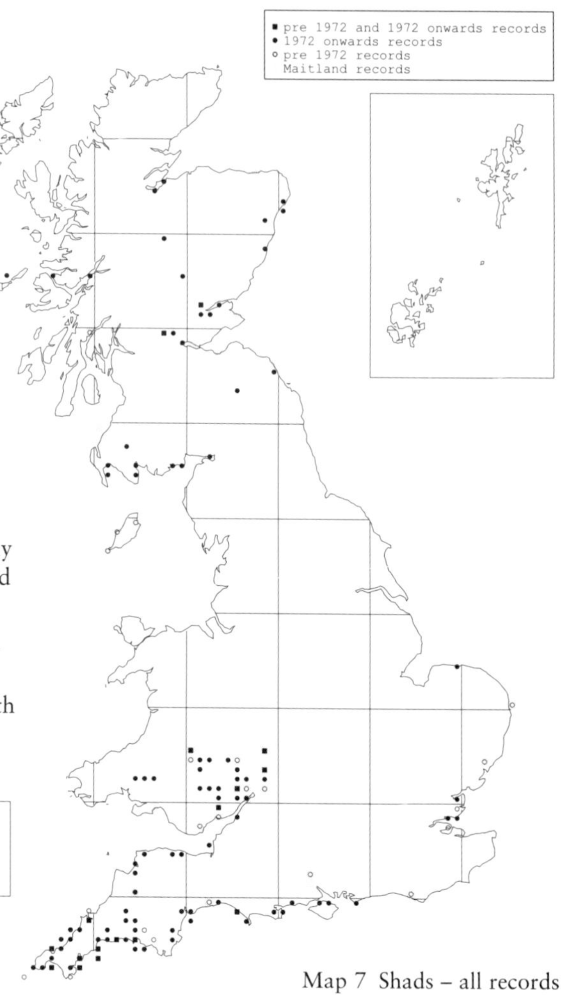

Map 7 Shads – all records

Description

- Both species have a deep, keeled body, laterally compressed, with large head and deeply forked tail.
- Lower jaw fits into a notch in upper jaw when mouth is closed.
- Back is dark blue, shading to gold on sides with silver below.
- With or, occasionally, without one or more dark spots along the sides from behind the gill cover to line of dorsal fin. Twaite shad usually have more spots than allis shad.
- Reliable identification requires the number of gill rakers to be counted.

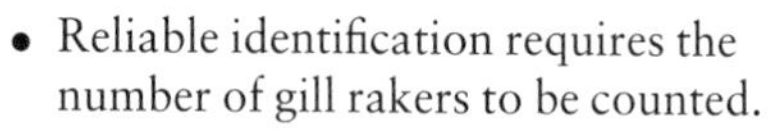

Size

Maximum adult length of twaite shad normally 40–45cm and allis shad slightly larger at 45–50cm. Males are normally smaller than females.

Biology and behaviour

Both species are anadromous, but allis shad normally spawn only once (though about five per cent spawn again) whereas twaite shad regularly spawn more than once. In Britain, male twaite shad mature at between three and four years old and allis shad at about four or five years. The females of both species mature a year later than the males. They migrate into the estuary in spring (April to May) when water temperatures reach 10°–12°C, entering the river in a series of waves, migrating upstream in daylight.

Although twaite shad have been reported spawning in tidal fresh water in the Elbe in Germany, in Britain both species appear to spawn in the river itself, as opposed to the estuary. The maximum distance that twaite shad migrate is around 180km, in the Wye. Both species spawn at night and spawning activity is brief lasting only a few seconds. The eggs, which are 2–4mm diameter, are broadcast into the water column and, being denser than water, sink to the bottom. They develop successfully between 15° and 25°C, incubation taking from 65 to over 160 hours, depending on temperature.

Fishes less than one year old migrate seaward during the autumn in the surface layers of the water column. In the Severn, the majority of twaite shad have left the estuary by the end of October; peak migration is associated with a decline in temperature with virtually none being caught once temperatures have fallen to below 9°C. Some year-old twaite shad reappear in the estuary in the spring and remain until the autumn, before again migrating seaward.

Juveniles of both species feed mainly on the immature stages of mayflies and flies, including non-biting midges (Chironomidae) and blackflies (Simulidae), in fresh water and on zooplankton in the estuary. At sea, Mysidacea (opossum shrimps) and fishes dominate the diet. The adults do not feed in fresh water.

Throughout their development, both species are similar in size although allis shad are usually slightly larger. Larvae measure from 4–12mm at hatching. When they migrate seaward in the autumn they have reached a mean length (to the tail fork) ranging from 4.5–6.7cm for allis shad and from 3.6–6.4cm for twaite shad.

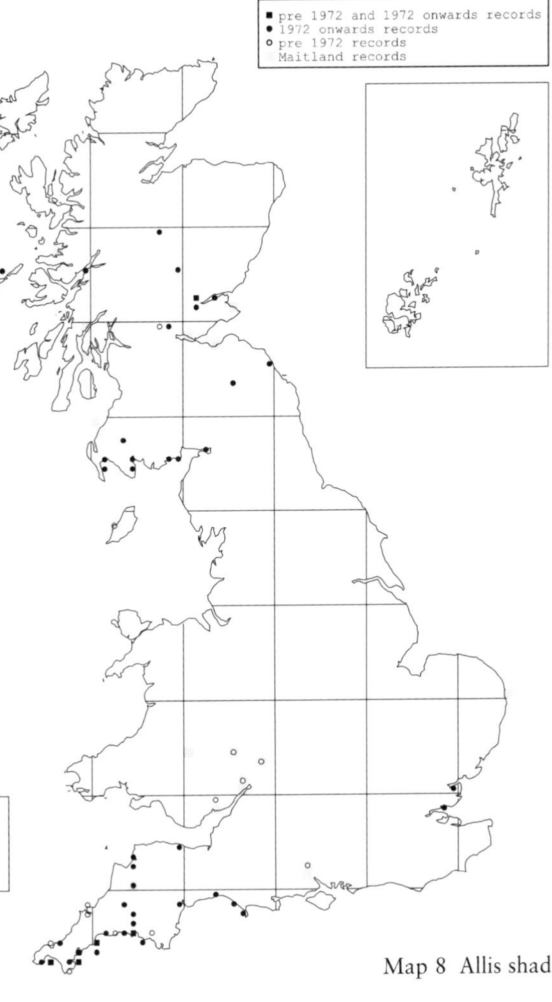

Map 8 Allis shad

Habitat

Very little is known about the marine habitat requirements of either species. At sea they are coastal in habit, occurring at depths between 10–150m. The spawning habitat of twaite shad in British rivers comprises fast-flowing, shallow areas of unconsolidated gravel/pebble and/or cobble substrate. The depth of water at spawning can range from 0.15–1.20m. Research on populations in the Elbe estuary, in Germany, showed that larvae prefer side-

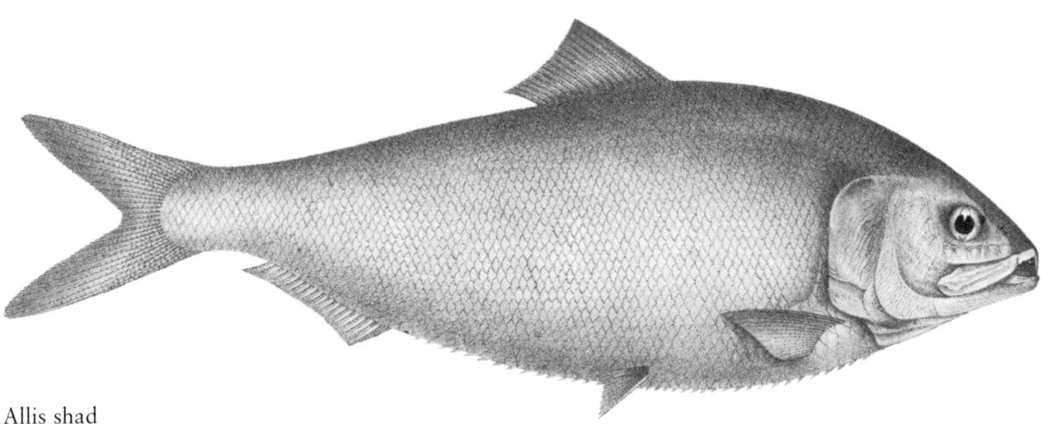
Allis shad

channels, which, because of the slower currents compared to the main channel, probably provide better nursery and feeding areas. It appears that the larvae actively avoid shallow areas close inshore.

Distribution in Britain

Both species have been reported from all around the coast of Britain, with the main concentration being in the Bristol Channel and around south-west England. Spawning populations of twaite shad are known to exist in four river systems, the Severn, Wye, Usk and Towy. Recent research strongly suggests the presence of spawning populations of both species in the vicinity of Wigtown Bay in south-west Scotland and of allis shad in the Tamar. There is good historical evidence to indicate that spawning populations of twaite shad once existed in the Thames and Trent, and of allis shad in the Severn. The two species are mapped separately (Maps 8 and 9) and together combined with undifferentiated records of 'shad' (Map 7).

World distribution

Both species are confined to Europe and north-west Africa, extending eastwards to Scandinavia and the Baltic Sea with reports from Iceland in the north and Morocco in the south. The allis shad also occurs in the western Mediterranean. Spawning populations of twaite shad are known to exist in Ireland, France, Germany, Portugal and Morocco, and allis shad in France and Portugal although the number of rivers supporting self-sustaining populations of either species has declined.

■ pre 1972 and 1972 onwards records
● 1972 onwards records
○ pre 1972 records
Maitland records

Map 9 Twaite shad

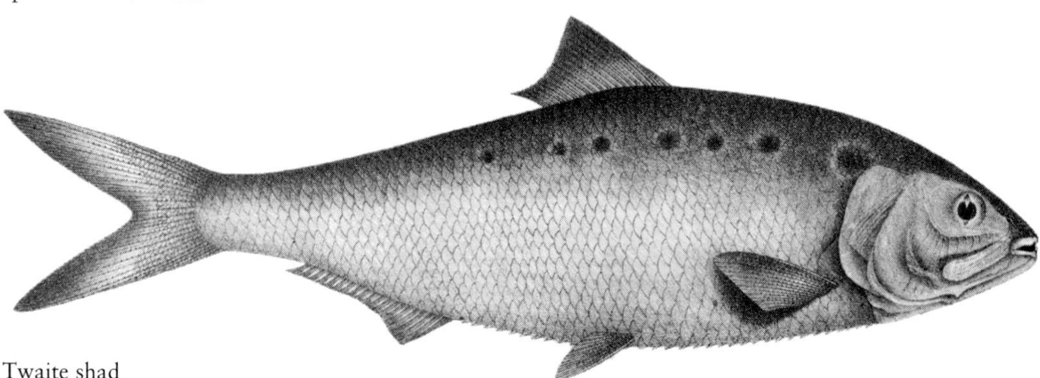

Twaite shad

Status

- Both species are protected under the Wildlife and Countryside Act 1981, the Habitats and Species Directive and the Bern Convention, and are listed as Priority Species under the UK Biodiversity Action Plan.

Shads and Man

In Britain, allis shad used to support a commercial fishery on the Severn in the nineteenth century, and twaite shad still support a small recreational fishery. Today, the main commercial fisheries for both species are in France with an annual catch of around 18 tonnes for twaite shad and between 500 and 600 tonnes for allis shad. In the past both species supported major fisheries elsewhere on the Continent, including the Rhine. Human influence on populations of both species in Europe has been mainly through the construction of barriers to migration and from pollution. Reference to historical information shows this particularly clearly for France: populations in many rivers, which used to be self-sustaining, are today extinct.

Further reading:

Baglinière & Elie, 2000; Billard, 1997; Keith, 1995.

Author: Miran Aprahamian, Environment Agency

Silver bream *Abramis bjoerkna*

Description

- Short, deep, laterally-flattened body with deeply-forked tail.
- Small scales with greyish silver coloration, pectoral and pelvic fins distinct pinkish-orange with pale grey tip.
- Small, slightly upturned mouth low on snout and large, pale golden eyes with large black pupils.
- Easily confused with young common bream and hybrids of roach and common bream.

Size

Typically grow to 20–25cm in length and 200–250g in weight, but can reach 35cm and about 1kg.

Biology and behaviour

Silver bream live in shoals, maturing at age three to five years when they are between 12 and 20cm. in length. They can live for up to ten years. They spawn in late spring and early summer, once water temperatures reach 14°–15°C, normally laying on submerged waterplants and occasionally on filamentous algae. A 20cm female will typically contain around 40,000 eggs which, when deposited, are pale yellow and around 1.7mm diameter. Silver bream feed principally on invertebrates including both planktonic and benthic animals, but they will take plant material

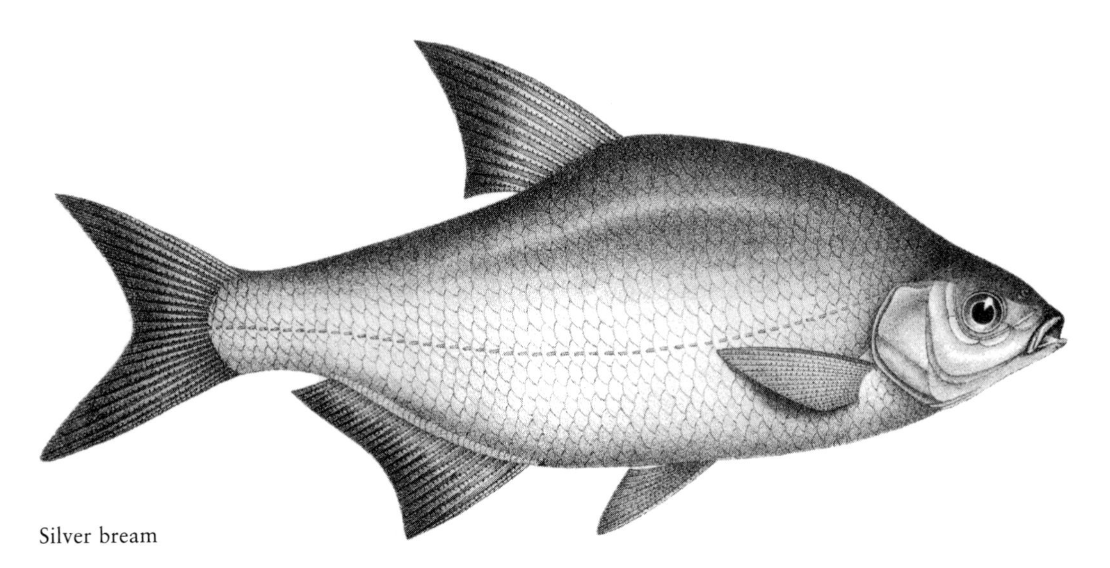

Silver bream

when other food is scarce. They feed most actively in the summer months although in Britain some feeding does occur during winter.

Habitat

Silver bream are typically found in turbid, slow-flowing, enriched lowland rivers and associated drains, canals and lakes, particularly those with steeply shelving margins, a silty substrate and some rooted waterplants. They can also be found in less turbulent areas in the middle reaches of larger rivers. Silver bream are able to tolerate some pollution and they thrive in some of the rivers now recovering from past histories of industrial and domestic pollution. They can also occur in brackish waters.

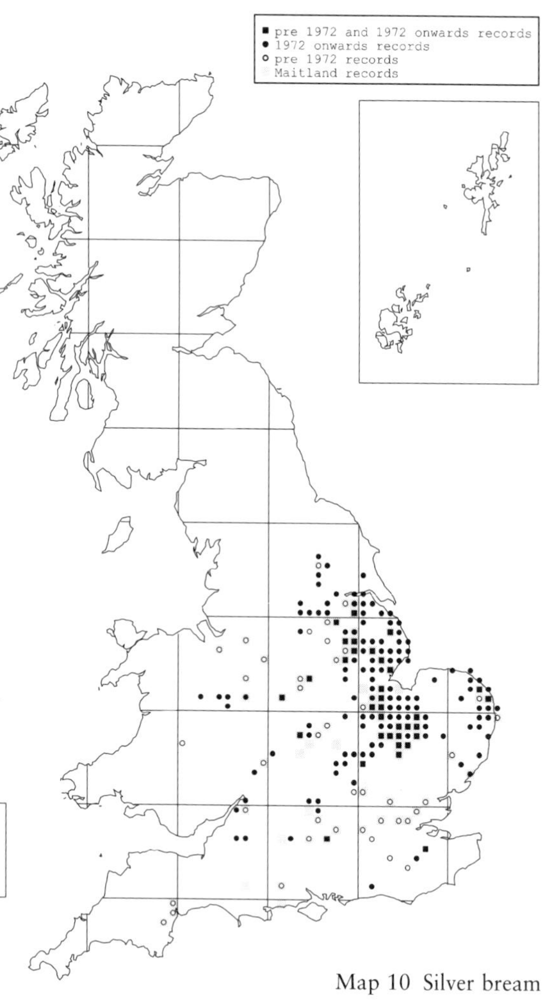

Map 10 Silver bream

Distribution in Britain

The silver bream has a more restricted natural distribution in Britain than many other cyprinid species and rarely forms a major component of fish communities. Originally confined to England in the large lowland rivers of the east Midlands and Yorkshire (the Ouse, Trent, Witham, Nene, Welland and Great Ouse and their tributaries) they are now found in many rivers, canals and still waters in the Midlands and south of England. The distribution mapped in 1972 indicates that silver bream were formerly recorded in other systems where they appear no longer to be present, especially in south-east England. However, this may have been due to earlier records of young common bream having been referred to by their colloquial name *silver bream*, and the possibility that true silver bream were never formerly present in these areas. In reality, the distribution of silver bream is probably now more widespread, due to deliberate or accidental inclusion in stocking programmes for angling, and as a result of water transfers between rivers. The species is absent from Scotland, south-west and north-west England, and Wales.

World distribution

The silver bream is common throughout the central part of continental Europe and eastwards into central Asia. It is absent from much of the Iberian, Italian and Balkan peninsulas, from the western and northern parts of Scandinavia, and from Iceland and Ireland.

Status

- Not considered rare or threatened, and not protected.

Hybrids and related species

The silver bream is related to other bream species in the genus *Abramis* and other cyprinids such as roach and rudd. In Britain, it hybridizes readily with common bream, roach and rudd.

Silver bream and Man

Due to its small size and lesser status in most fisheries, the silver bream is of little importance in the commercial fisheries of continental Europe and is not sought as a sport fish. However it has sometimes formed the mainstay of recreational angling catches in English rivers such as the Trent.

Author: Graeme Peirson, Environment Agency

Common bream
Abramis brama

Description

- Large, bottom-feeding species with deep, laterally compressed body and dark brown fins.
- Silvery flanks become darker and more golden-olive with age; white or creamy belly.
- Dorsal fin set well back, long anal fin extends almost from middle of belly to deeply forked tail.

Size

Maximum length 80cm and weight up to 9kg.

Biology and behaviour

Common bream reach maturity between the fourth and sixth year of life and may live up to 20 years. Spawning usually takes place in May or June among dense vegetation in very shallow water, at temperatures between 12° and 20°C, and may be repeated once or twice at weekly intervals. During spawning males have numerous white or yellow tubercles on the head and front of the body and occupy small territories, defending them against other males. Females can lay between 150,000 and 300,000 eggs per kilogram of body weight.

Common bream normally swim in shoals, behaviour that begins in the larval stage and continues throughout their development to mature adults. Old individuals tend to be solitary or they can form small groups. Adults are capable of migrating long distances (up to 60km) and in parts of their range have been described as semi-migratory, with some populations moving between brackish and fresh waters. Juveniles feed mainly on zooplankton in the water column, but the adults shift towards bottom feeding on invertebrates, with a preference for small crustaceans and the larvae of non-biting midges (Chironomidae).

Habitat

Common bream are characteristic of nutrient rich, lowland lakes and slow-flowing rivers with a clay or muddy bottom. They are able to survive in waters with some pollution and can tolerate very oxygen levels. They have an optimum temperature range for growth between 20° and 28°C, and an upper lethal limit of 32°C. Diversity of habitat is less critical for common bream than

Common bream

for many other cyprinids and they will flourish in relatively homogeneous environments.

Distribution in Britain

An indigenous species, it is fairly widespread and common, particularly in slow-flowing rivers, canals and nutrient rich lakes. They are found throughout England and are particularly dominant in the lowland watercourses of East Anglia and Cheshire, and in the Trent and the Yorkshire Ouse. Where they occur, common bream can often be found in large numbers and now appear to be more widespread within their traditional range

Common bream are generally absent from Scotland, much of Wales and the extreme southwest of England because suitable natural habitat is not available, but comparison with the distribution in 1972 shows some increases in these areas. Records from south Wales include several large ornamental ponds, for example in urban parks, but the recently established localities in south-west Scotland are mainly large lakes and reservoirs in remote areas. The recent records from Devon are due to the construction of purpose-built still waters for angling. The apparent decline of common bream in some areas is almost certainly due to an absence of recent records and not because the species is now absent or in general decline.

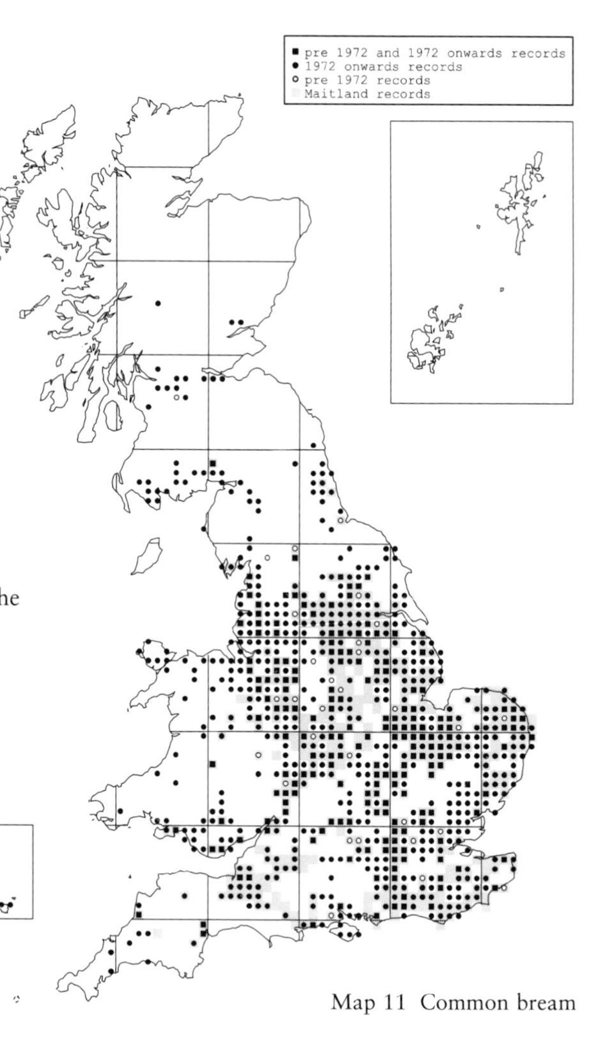

Map 11 Common bream

World distribution

Common bream occur in both fresh and brackish waters throughout continental Europe. In fresh water they are found as far south as the Alps and Pyrenees, throughout the Balkans to Northern Turkey and the Caucasus, and east to the Urals and the central Asian steppes. The range extends north into Scandinavia and they are found in the brackish waters of the Baltic.

Status

Not threatened or protected in Britain or the rest of Europe.

Hybrids and related species

Common bream commonly hybridize with roach, the progeny often being fertile, and occasionally with rudd and silver bream. Closest relatives are other species of the genus *Abramis*, such as silver bream and two central European species, white-eye bream *A. sapa* (Pallas) and zope *A. ballerus* (Linnaeus).

Common bream and Man

Common bream are popular as sport fish, particularly with competition anglers where the species can account for large catches. In eastern Europe the species is considered one of the most important commercial fishes for human consumption. The flesh of fishes over one kilogram in weight is soft but bony, and with a not unpleasant but indistinct flavour.

Further reading:

Backiel & Zawiska, 1968; Lammens *et al.*, 1987; Langford, 1981; Mann, 1996; Whealan, 1983.

Author: Jim Lyons, Environment Agency

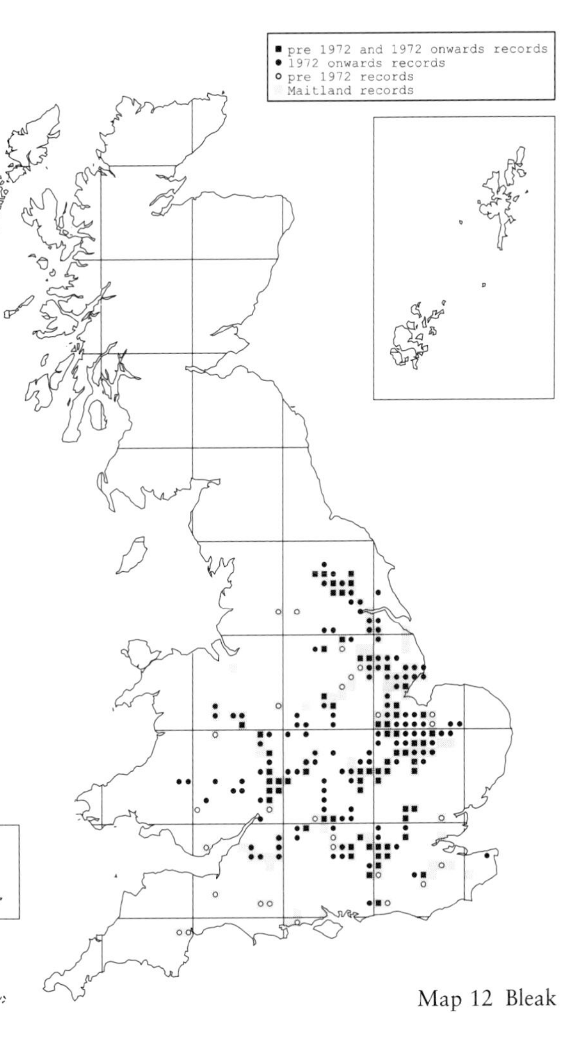

Map 12 Bleak

Bleak *Alburnus alburnus*

Description

- Very slender, laterally flattened body.
- Small delicate scales give body bright, silvery appearance, with pale green on upper flanks and back, and pale yellow fins.
- Head with comparatively large, upturned mouth and protruding lower jaw and large eyes.

Size

Can grow to 20cm, but often only 15cm in length, and to 35g in weight.

Biology and behaviour

The bleak is a small, short-lived species, rarely living more than five or six years and in many waters only to three or four years. They live in large shoals and feed mainly on planktonic animals and small insects at the surface and in the water column, only occasionally from the substrate. Bleak are active chiefly in summer and tend to form dense, localized aggregations in winter, although

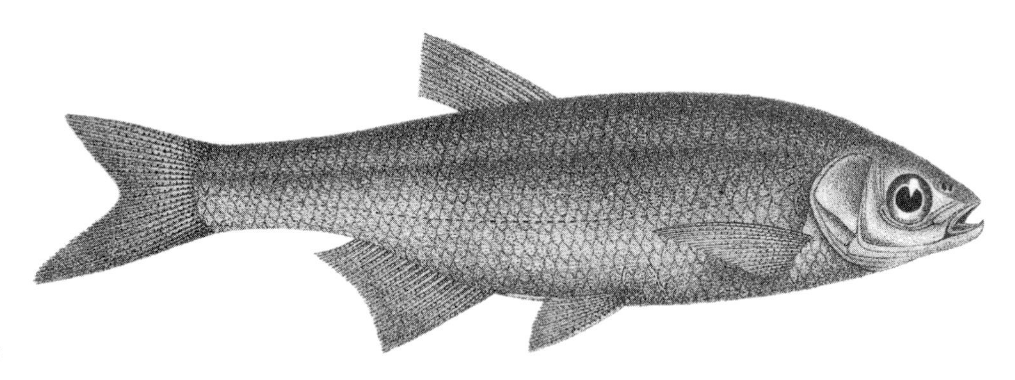

Bleak

they characteristically disperse during winter flood conditions, when they can often be observed feeding at the surface in slack water. Bleak mature aged two or three years old and a 15cm female will produce around 5,000 eggs. Spawning takes place in spring and early summer on gravel, stones and associated waterplants after water temperatures have reached 15°C. Eggs are around 1.5mm diameter and pale yellow or cream and are released in batches over a period of several days.

Habitat

Bleak favour enriched, but fairly clean, flowing water in the middle and lower reaches of larger rivers, and are also found in associated drains, canals and backwaters. In continental Europe they also form self-sustaining populations in lakes.

Distribution in Britain

The natural distribution of bleak in Britain is thought to have been the river systems draining into the Humber and the Wash, and also the river Thames. They were introduced to other river systems, probably during the eighteenth or nineteenth centuries, possibly initially for use as bait for pike or other species. Bleak are now abundant in the Wye, Severn, Warwickshire Avon and Bristol Avon and associated canals and drains, but they are absent from the East Anglian rivers that flow eastwards to the North Sea, and from much of the extreme south of England. They are also absent from the north-east, north-west and south-west of England, west Wales and Scotland. Apparent contractions in range in the west of England and north-west Midlands since 1972 may be due to earlier introductions of bleak as bait fish which have not survived. However, the absence of records from the middle Severn and lower Trent probably underestimates its present-day range in these areas.

World distribution

Bleak are widespread and common throughout continental Europe west of the Urals, with the exception of the Iberian peninsula (although recent accidental introductions there have been reported), Italy, western and northern Scandinavia, Iceland and Ireland.

Status

- Not considered rare or threatened, and not protected, but may have declined in abundance in Britain.

Hybrids and related species

The bleak is one of several similar species in the genus *Alburnus*, although *A. alburnus* is the only species found in Britain and most of the rest of Europe. It is closely related to other common cyprinids and may hybridize with chub, roach, dace and silver bream.

Bleak and Man

Despite its small size the bleak used to be important as food and commercially, probably because it was so abundant in many fisheries. Its flesh is tasty and would have been a major component of freshwater 'whitebait' dishes; also, the tiny scales were used to make artificial pearls and nail varnish. For recreational and particularly for competitive anglers the attraction of bleak is that under certain conditions they can be induced into feeding frenzies at the surface when large numbers can be caught in a short space of time. During a contest in the early 1980s, a French angler caught 515 bleak in one hour!

Author: Graeme Peirson, Environment Agency

Barbel *Barbus barbus*

Description

- Long streamlined body covered in bronze and golden scales.
- Prominent, pointed and concave dorsal fin, large pectoral and pelvic fins. Fins reddish-brown colour.
- Four distinct barbels set around downward protruding mouth.

Size

Normally up to 60cm long but exceptionally to 90cm; rarely exceed 6kg in weight, but specimens of over 7kg have been caught in Britain.

Biology and behaviour

In suitable conditions barbel may live for up to 25 years but few live beyond ten. Males typically mature after three years and females after five years. They spawn over shallow, gravelly riffles during June and July, when water temperatures reach 18°C; a typical female will produce up to 50,000 eggs. Barbel generally remain within a limited home range, provided suitable spawning habitat and feeding areas are nearby, but they are able to migrate for long distances and negotiate low weirs and fish passes designed for salmon. They tend to shoal in small 'family' groups remaining close to the river-bed where they forage predominantly for crustaceans and small fishes. Although not overtly aggressive, they are highly competitive with other bottom-feeding species.

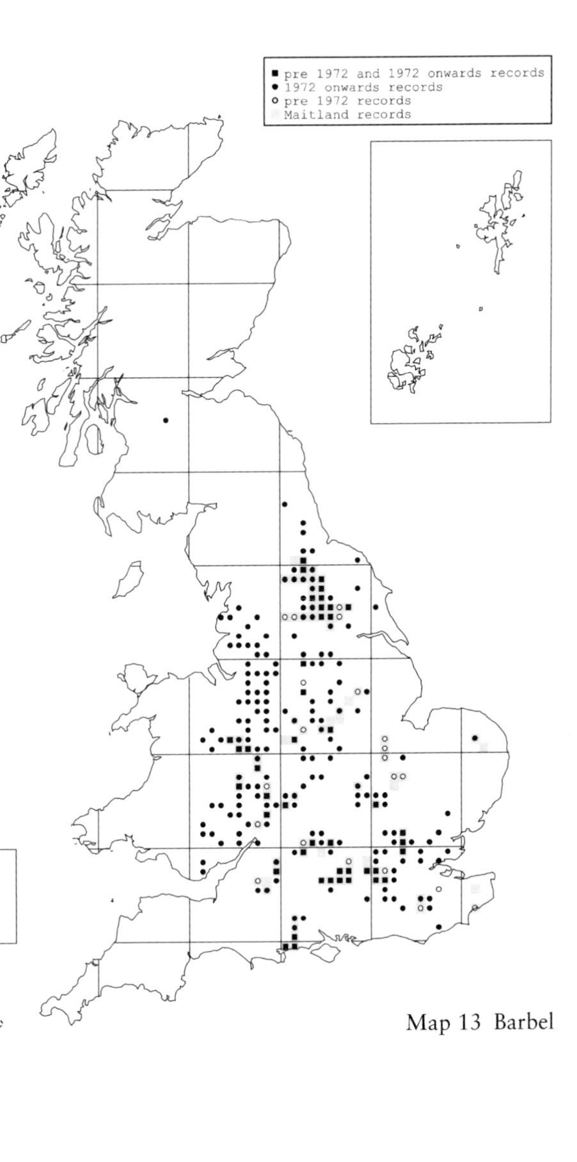

Map 13 Barbel

Habitat

Barbel are capable of living in many different habitats, including still waters, although they will reproduce and thrive only in fast-flowing rivers and streams. Soon after hatching, the fry often congregate in shallow, slow-flowing, marginal areas where they spend the first few months of life. As they grow, they move away from the margins to mainstream habitats more typically associated with adult barbel. These include areas of shallow, fast-flowing water, particularly with clean gravel substrate and prolific weed growth. They seek cover in weed during daylight hours and forage in open water at night. In the absence of weed they will seek refuge in deep water or shaded areas beneath overhanging marginal vegetation. Barbel do best in a plentiful supply of good quality water with temperature ranges from 10° to 18°C.

Barbel

Distribution in Britain

It has been suggested that barbel used to be confined to a few rivers along the eastern seaboard of England, being remnants of more widespread populations in the days when Britain was connected to mainland Europe in the immediately post-glacial period. As angling became progressively more popular during the late nineteenth century barbel were recognized for their high sporting potential and were subsequently relocated to other rivers around the country. For example, from sketchy records, it appears that during the 1890s the first barbel to reach the River Stour in Dorset were transported in beer barrels from the River Kennet in Wiltshire, a tributary of the River Thames. Currently, barbel can be found in many rivers across England, except in the south west, and in a few locations in Wales. They do not occur naturally in Scotland, but have been introduced to and are now breeding in the River Clyde.

World distribution

Barbel are found across most of continental Europe, with the exception of the Iberian, Italian and extreme Balkan peninsulas, and they are absent from Scandinavia and Ireland.

Status

- Populations fluctuate, but not threatened.

Hybrids and related species

- No hybrids known in Britain.
- Several other species of *Barbus* occur in southern Europe.

Barbel and Man

Barbel are no longer thought of as a source of food in Britain, but they are still eaten in Eastern Europe and occasionally by some ethnic groups in Britain. They are highly sought after as a sport fish and particularly attract anglers who dedicate much time, effort and money in the pursuit of large specimens.

Author: Matt Carter, Environment Agency

Goldfish *Carassius auratus*

Description

- Distinctive, rotund fish, normally orange coloured, sometimes with black, white or red patches.
- Feral (wild) goldfish usually dull brown colour with silvery-gold belly.
- Dorsal fin long, straight/concave with strongly serrated spine, pelvic fins usually pale, tail deeply forked. Some ornamental forms with elongated fins.
- Commonly 28 scales on strong continuous lateral line.

Size

Rarely more than 30cm in length or above 1kg in weight.

Biology and behaviour

Goldfish usually spawn in June or July, when water temperatures exceed 20°C. The small, sticky, yellow eggs attach to plants near the water surface in thick weed beds. Hatching usually takes place between three and eight days. Goldfish normally mature at three or four years. Up to 400,000 eggs per fish can be laid, dependent on the size of the female. Goldfish are capable of gynogenesis in the absence of males; the sperm of other carp species can stimulate egg division to develop young fishes genetically identical to their mother. This can result in entirely female populations. Hybrids between other species and goldfish (see below) are also likely to be capable of gynogenesis, which will lead to further competition with pure line parents. Goldfish are usually found in small groups. Juvenile fishes feed on zooplankton, but as they grow, they feed mainly on the bottom, on small molluscs, worms, crustaceans and insects (such as chironomid larvae), and plant material.

Habitat

Goldfish seem to prefer small, rich ponds and lakes with abundant macrophytes, although they are also recorded in smaller numbers in slow-flowing, lowland rivers. They are extremely hardy and very tolerant of low levels of dissolved oxygen, and can tolerate temperatures of near 0° to 40°C. Goldfish appear to survive in most places where they are stocked and can exist in very dense populations causing habitat degradation through suspension of silt.

Distribution in Britain

Goldfish were probably introduced to Britain in the late seventeenth century and are widely naturalized, being found in ornamental ponds in parks and on large estates and, increasingly, in still waters used for commercial sport fisheries. Goldfish are commonly released into the wild as unwanted pets, for example from overstocked garden ponds, and more recently also through mass stocking of sport fisheries. A population of goldfish in the Forth–Clyde Canal disappeared when an outfall of warm water from a factory ceased. Compared with the distribution recorded in 1972, the map shows an increase in occurrences throughout southern Britain. This recorded distribution under-represents the probable range of the species because of the lack of records from still waters (especially ornamental ponds) and possibly also through misidentification with crucian carp.

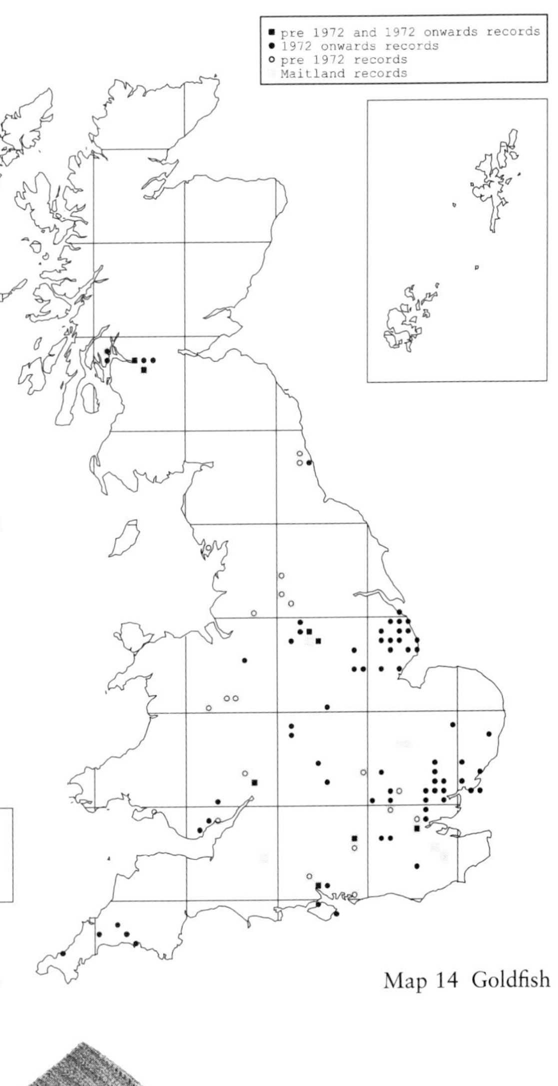

Map 14 Goldfish

World distribution

Goldfish are native to eastern Asia, but have become naturalized throughout lowland Europe and in many other continents.

Status

- Becoming increasingly common in the wild.
- Goldfish represent a significant threat to British ecosystems through competition, hybridization and the spread of disease.

Goldfish

Hybrids and related species

- Goldfish hybridize readily with common carp, often producing offspring capable of further reproduction, and also are believed to hybridize commonly with crucian carp in Britain and possibly wherever their ranges overlap.
- The taxonomic status of the species described as *C. auratus* is subject to debate. Some authors consider goldfish to be a domesticated form of the wild gibel carp, alleged to be native to eastern Europe and Asia, whereas others regard gibel carp as a distinct subspecies, *C. auratus gibelio* (Bloch). Gibel carp (also known as Prussian carp) may simply be feral goldfish or possibly hybrids between goldfish and crucian carp. Despite intensive studies, no further resolution of this issue seems possible without the use of molecular biological techniques.

Goldfish and Man

Goldfish are extensively farmed for the international aquaria trade. Selective breeding, originally started in Japan and China, has created fishes with a huge array of different coloured, often distorted forms including spherical bodies, bubble eyes, lion heads and fan tails. Thus, goldfish vary in shape and colour depending on their origin and closeness to an ornamental strain. Identification of feral goldfish is difficult due to morphological differences, and they are often misidentified as crucian carp. An increasing number of commercially operated sport fisheries provide goldfish as a target species.

Author: Philip Bolton, Environment Agency

Crucian carp
Carassius carassius

Description

- Typically very deep bodied and laterally compressed, but body depth can vary considerably. 'Stunted' populations are typified by their small size and thin body-shape.
- Head small and round; mouth with up-turned lip.
- Rusty bronze coloured flanks with golden yellow/orange belly.
- Dorsal fin long, convex with slightly serrated spine, anal and pelvic fins often orange with black tip, tail blunt.
- Commonly 33 scales on incomplete lateral line, usually fading towards tail.

Size

Average maximum length 20–30cm but 'stunted' crucian carp are smaller.

Biology and behaviour

Spawning usually takes place in May or June, when water temperatures exceed 14°C, near the water surface where one or more males actively chase the ripe female. The size of the female determines the number of eggs laid, but up to 300,000 eggs per female is possible. The small, sticky, yellow eggs attach to plants in thick weed beds. Depending on temperature, hatching usually takes place in five to seven days. Crucian carp are normally found in small groups. Juveniles feed mainly on zooplankton, but as they grow they move to feed mainly on the bottom, on small molluscs, worms, various insects (such as chironomid larvae), crustaceans and some plant material. Crucian carp usually mature at three to four years and live up to ten years, although individuals may live significantly longer.

Habitat

Crucian carp tend to prefer small, rich ponds and lakes with abundant macrophytes in lowland areas. They are very hardy: tolerant of low levels of dissolved oxygen, can survive in temperatures of near 0° to 38°C and can live in acidic waters as low as pH4. Very dense, stunted populations are recorded in Scandinavia, where they live in monocultures sealed below thick ice and snow for up to six months of the year. This water reaches near freezing temperatures and is almost without oxygen. Here they thrive and spawn repeatedly during the summer months. The typical larger, deep-bodied crucian carp tend to come from waters with less extreme conditions where there are other species, including predators, and competition is between species rather than intra-specific.

Distribution in Britain

Crucian carp are considered by some authors to be native to south-east England. Others regard the species as having been introduced to this area, possibly in mediaeval times. If crucian carp are native, they have also been introduced to other areas, but are confined mainly to lowland waters. Comparison with the distribution mapped in 1972 shows that the species has spread further north and west across Britain. This may give a misleading impression due to the difficulties in identification of crucian carp and some records may be of misidentified feral goldfish and goldfish/crucian carp hybrids. The range is likely to have been extended due to the increase in transfers for commercial sport fisheries, although this has probably also had the effect of introducing juvenile feral goldfish (accidentally mixed with juvenile crucian carp) into otherwise pure stocks of crucian carp, resulting in both hybridization and inter-specific competition.

■ pre 1972 and 1972 onwards records
● 1972 onwards records
○ pre 1972 records
Maitland records

Map 15 Crucian carp

Crucian carp

World distribution

Crucian carp are native to eastern, central Europe and much of central Asia. Some authors regard them also to be native in parts of western Europe.

Status

- As either a native species, or a long-established introduction, there is concern that crucian carp are threatened by habitat degradation and competition, hybridization and infection with an introduced carp tapeworm.

Hybrids and related species

- Hybridizes readily with common carp, producing fertile offspring capable of reproduction.
- It is believed that crucian carp commonly hybridize with goldfish in Britain and also possibly wherever their ranges overlap.

Crucian carp and Man

The accurate identification of crucian carp is a problem worldwide due to morphological similarities with feral goldfish and their hybrids. This is further complicated because, in Eastern Europe and Asia, goldfish (*C. auratus* and presumed subspecies) are frequently referred to as crucian carp. In 1998 the British rod-caught record for crucian carp was reviewed and it was decided that previously accepted record fishes from 1950 or earlier were actually feral goldfish or goldfish hybrids. A new qualifying weight of 0.91kg for crucian carp was then introduced. A crucian carp captured in Hampshire weighing 2.04kg holds the current British rod-caught record.

The crucian carp is under threat from man's activities. It is now unusual to find a population of pure crucian carp due to introductions of common carp and goldfish to waters with crucian carp. Hybridization with these species threatens pure populations of crucian carp in Britain and hybrids are often capable of further reproduction and are able to produce fertile offspring. Competition, particularly with common carp, is also a significant threat to crucian carp. The advent and popularity of the commercial carp fishery where high stock densities of carp are introduced into lakes and ponds usually results in degradation of the habitat and severely impacts on native crucian carp populations. The Asian tapeworm *Bothriocephalus acheilognathi* Yamaguti, believed to have been introduced into Britain with imported grass carp, is also a threat as crucian carp populations at several sites are known to have been adversely affected.

Author: Philip Bolton, Environment Agency

Grass carp
Ctenopharyngodon idella

Description

- Shape more like a chub than a carp: slender cylindrical body with large broad head, small eyes and large, wide mouth.
- All fins rounded in outline; dorsal and anal fins short based, with front edges of dorsal and pelvic fins in line.
- Scales moderately large. Conspicuous ridges on opercula.
- Back dark olive; sides and belly grey, tinged with gold.

Size

Adults can reach up to 1.25m in length and 18kg in weight.

Biology and behaviour

Grass carp are unable to breed in north-west Europe, as they require quite fast-flowing water with temperatures of 23°–25°C for approximately nine weeks to do so successfully. In warmer climates they breed in spring: the pelagic eggs are distributed by water flow and hatch quickly. Young grass carp feed on small invertebrates, but the adults feed only on aquatic plants and are capable of altering ecosystems significantly.

Habitat

The natural range of the grass carp is in the middle and lower reaches of the River Amur in south-eastern Russia, but in Britain it has been acclimated (but not naturalized) in rich lakes and slow-flowing rivers.

Distribution in Britain

Originally introduced to Britain in the 1960s as farmed stock from Hungary, grass carp were brought into enclosed waters, mainly in the Fens, to control the growth of macrophytes. Although they cannot breed in the British climate, they are now widely distributed in England and Wales. The species was not recorded in the 1972 maps.

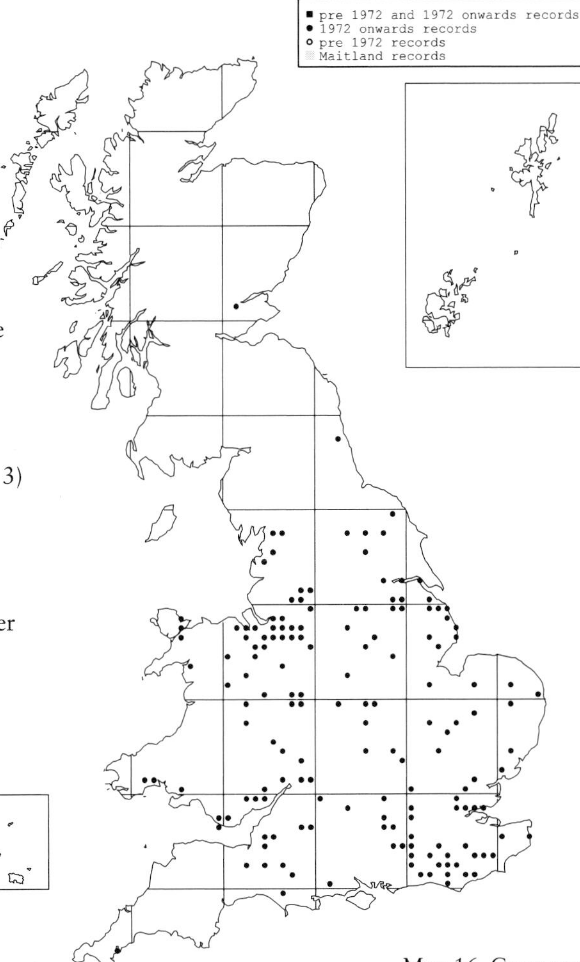

Map 16 Grass carp

World distribution

Native to the River Amur, they are found in eastern Russia and China and have been introduced into many parts of central Europe and North America.

Status

- Listed on the Prohibition of Keeping or Release of Live Fish Orders (see Appendix 3) as a species for which release to the wild is not permitted without a licence.

Hybrid and related species

- In warm climates often hybridize with silver carp *Hypophthalmichthys molitrix*.

Grass carp and Man

Although grass carp were introduced to control excessive weed growth in eutrophic waters, more recently they have become popular with anglers. In Asia they are farmed for food. Small specimens can even be kept in aquaria.

Authors: Sarah Chare & Robin Musk, Environment Agency

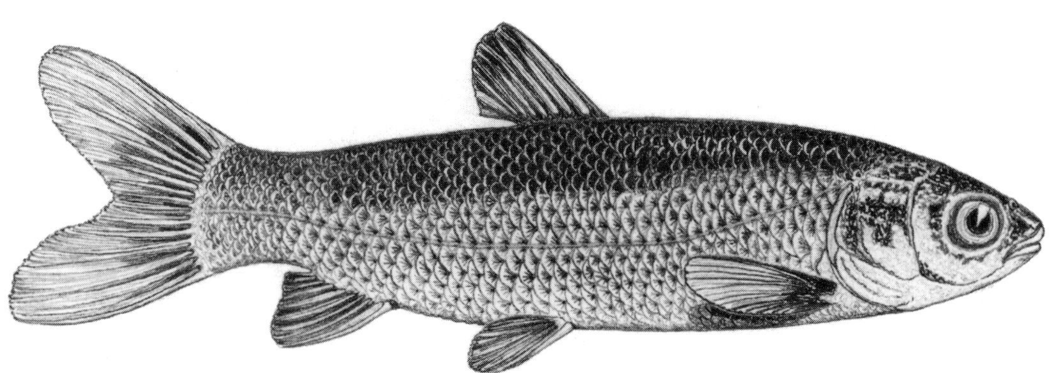

Grass carp

Common carp *Cyprinus carpio*

Description

- Body usually thickset to rounded; head with characteristic large, protrusile, rubbery lips with two pairs of barbels, one short and one long.
- Dorsal fin long, concave with strongly serrated spine; powerful tail.
- Four different forms:
  - common carp, with an even spread of large scales;
  - mirror carp, lightly covered in rows or sporadic, unevenly placed, oversized scales;
  - leather carp, almost totally without scales;
  - koi carp, highly coloured, ornamental forms.
- Depending on form and habitat, colour ranges from silvery to golden to chestnut brown flanks, fading to creamy white belly.

Size

Rarely more than 1m in length or 6kg in weight, but occasionally much greater (see below under 'Carp and Man').

Biology and behaviour

Carp are warm-water species and are considered to require water temperatures of at least 18°C to spawn; in Britain these are normally reached in May or June. Spawning is usually in the shallow margins of a lake or river where the sticky eggs are deposited on weed, submerged roots or other firm substrates. Spawning usually takes place at dawn and is energetic, often with vigorous splashing in the weed beds as the female lays eggs whilst being chased by one or more males; the female may even be lifted out of the water by the eager males. Common carp are very fecund and a 5kg female may produce up to a million eggs. The temperature requirement means that there are many waters in Britain where carp never spawn successfully and which support only stocked populations.

Common carp are social and are generally found in small shoals. They are omnivorous and usually feed on the bottom, preferring weedy, silty margins where food is plentiful. Newly hatched carp larvae feed on single celled protozoa and algae. After several days, the fry begin to feed on larger items such as cladocerans and copepods. Having reached a length of 200mm, carp feed mainly on plant matter and larger invertebrates. Carp commonly mature at three to four years and live on average for 10–20 years, but can live for over 50 years.

Habitat

Carp typically occur in lowland lakes and rivers where the waters are rich with abundant vegetation to provide food, shelter and spawning substrates. Where conditions are suitable, carp often alter the habitat: waters containing high densities of carp usually change in character as the population expands, often becoming turbid due to disturbed silt inhibiting the growth of submerged macrophytes. Carp are adaptable and can be found in most freshwater conditions, from cold, upland streams and large, wind-swept, gravel pits to warm, nutrient rich, farm ponds. They will reproduce in many different environ-

Common carp

ments if the water is warm enough and can tolerate water temperatures of near freezing to 40°C, brackish conditions, low oxygen content and relatively poor water quality.

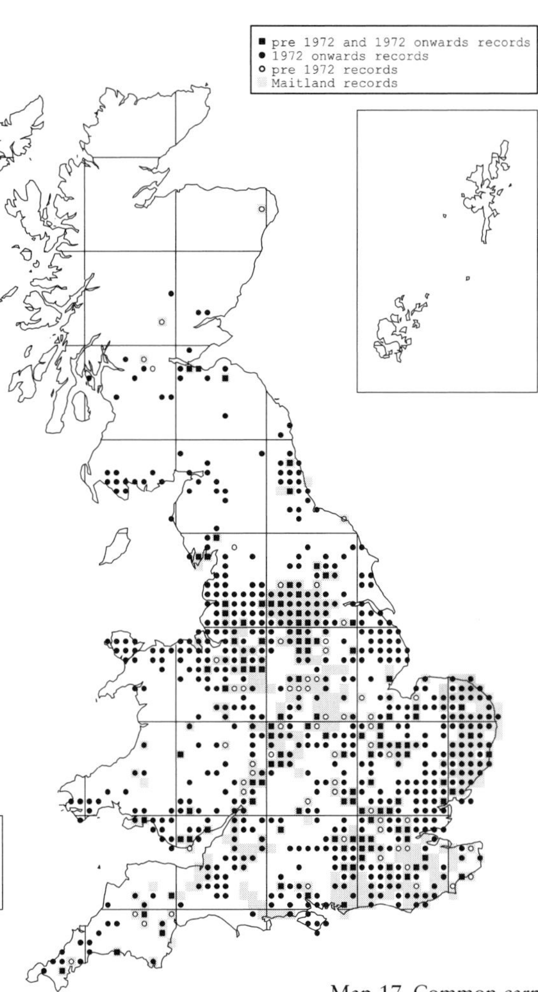

Map 17 Common carp

Distribution in Britain

Carp were introduced into Britain sometime before AD 1500. They are now found widely throughout England and Wales, except areas of high ground, and in only a few areas in southern Scotland. Compared with the distribution maps of 1972, the range of carp appears to have expanded northwards. This probably reflects a genuine increase in carp introductions for sport fisheries. However, the distribution mapped here under-represents the scale of these introductions because there was no integrated system for recording them. There are areas where they were mapped in 1972 from which recent records of carp have not been received, probably reflecting the absence of a consistent system for recording fish stocking and the fact that relatively few data covered still waters.

World distribution

It is likely that the natural range of carp is limited to Asia, but this natural range has been greatly extended by introductions worldwide including across Europe, Turkey, Israel, Africa, India, Australasia and North America. This list reflects the popularity of carp as a food source, as a sporting fish and as an ornamental species, and its ability to colonize.

Status

- Not threatened in Britain, the rest of Europe or globally.
- Considered to be a pest species in some countries.

Hybrids and related species

- Readily hybridizes with crucian carp and goldfish producing fertile offspring.

Carp and Man

Carp are one of the most important freshwater fishes in the world in terms of economics. Depending on where they are in the world, they are either loved as food or loathed as a pest. The practice of culturing common carp was described in China by Fan Li in 473 BC and they have been farmed ever since for food, for sport and for their aesthetic appeal or as pets. It is an ideal farm species being hardy, non-aggressive, fast growing and well flavoured with firm white flesh. The original or 'wild' carp is a lean, streamlined fish with a full complement of even scales but over thousands of years the 'wild' carp has been domesticated by fish farming to produce the king carp. This is a fast growing strain of scaleless (leather) or lightly scaled (mirror) carp produced primarily for the table. In Japan, carp have been selectively bred to produce highly coloured varieties, the Nishikigoi or 'koi' carp.

Carp were originally introduced to Britain in the Middle Ages, probably as a source of food and although they are rarely eaten here, carp are still an important food source in many other countries. In the Czech Republic, for example, where carp ponds are widespread, a special carp dish is the traditional Christmas fare. The fortunes of carp in Britain have changed as they are now one of our most popular sport fishes. Unlike popular game fishes (trout and salmon) they are almost invariably returned live to the water. The size of the wild British carp is increasing and many of the larger fishes are now named. 'Mary', the British record rod-caught carp, weighing 25.6kg, was recently eclipsed by a mirror carp known as 'Two Tone' weighing in at a huge 27.7kg!

The carp is classed as vermin in North America, Africa and Australasia due to its ability to colonize and rapidly alter ecosystems by its feeding habits. Many different task forces exist in these continents with the sole intention of eradicating or controlling carp numbers in the wild.

Further reading:
Froufe *et al.*, 2002

Author: Philip Bolton, Environment Agency

Gudgeon *Gobio gobio*

Description

- Body streamlined, with chisel-shaped head flattened underneath, similar to a miniature barbel.
- Colour iridescent, silvery purple-blue, with darker brown patches along lateral line and brown spots on dorsal fin and tail.
- Long, narrow caudal peduncle.
- Mouth small, profiled downwards and protrusile, with well-developed single barbel at each corner of mouth.

Size

Length 7–12cm, maximum 20cm; weight about 100g up to a maximum of 225g.

Biology and behaviour

Both sexes are mature at two or three years of age when sexually mature males develop pale bumps (tubercles) on the head and shoulders. Females may carry 800–3000 eggs, which are laid in several batches. Spawning is between May and June and occurs over a period of several days on gravels and macrophytes in shallow water.

Gudgeon often form shoals, particularly in warm, shallow water, and feed mainly on the bottom, on molluscs, insect larvae and crustaceans in or on the substrate. They are able to communicate in shoals by a series of 'squeaks'. Gudgeon are normally active during daylight, but can switch to semi-nocturnal activity if significantly threatened by predators such as large trout, pike and goosander *Mergus merganser*.

Habitat

Gudgeon favour fast-flowing streams and rivers with a sand and/or gravel substrate. However, they are also found in still and slow-flowing waters, such as the Norfolk Broads, and are abundant in much of the canal system of Britain.

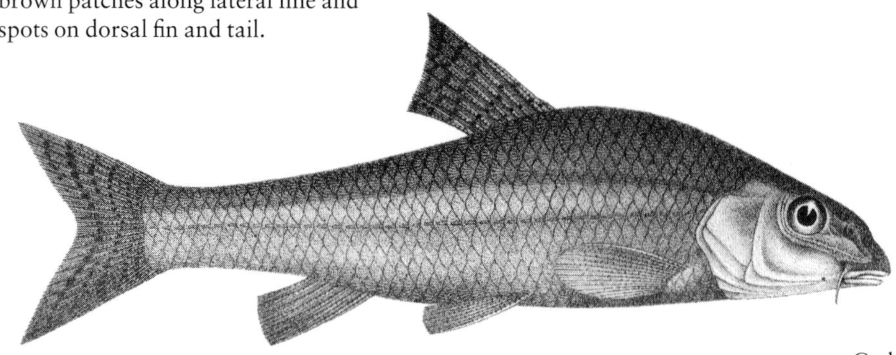

Gudgeon

Distribution in Britain

Gudgeon are found throughout most of the southern half of Britain but are absent from south-west England, parts of western Wales, much of Scotland, and all offshore islands. Comparison with the distribution recorded in 1972 reveals an apparent slight spread, westward in the Welsh borders and Wales, and northwards in England.

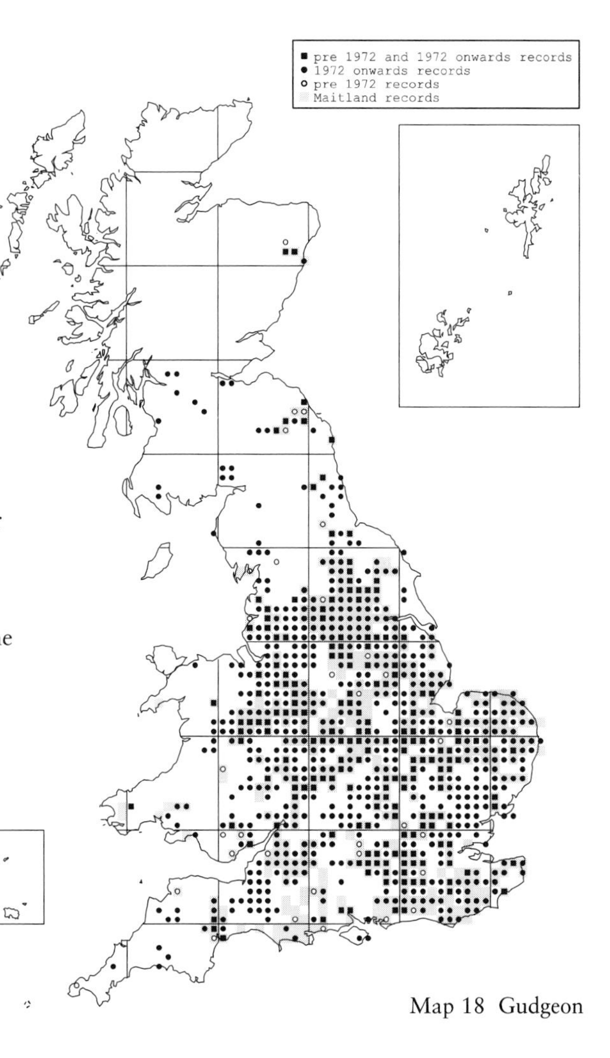

Map 18 Gudgeon

World distribution

Gudgeon are widely distributed throughout Eurasia from Ireland to Asiatic Russia, but are absent from northern Scandinavia, much of Spain, central and southern Italy and southern Greece. Several related species occur in Europe (see below).

Status

- Not threatened or protected in Britain or the rest of Europe.

Hybrids and related species

- No hybrids known in Britain.
- Gudgeons, *Gobio* species, comprise a distinctive group of the family Cyprinidae, which includes minnow and the carps. There are eight species worldwide, most of which are found in Asia. There are five species in Europe with several, such as the Danubian long-barbel gudgeon *Gobio uranoscopus* (Agassiz), confined to the Danube and Dniester river basins.

Gudgeon and Man

Although gudgeon are eaten in France, as with most freshwater fishes they are not eaten in Britain despite being reportedly delicious. During fishing matches they are either the bane of anglers or a boon, as they can be caught in large numbers. In the nineteenth century, gravel shallows were raked over to attract feeding gudgeon for angling.

Author: Andrew M. Hindes, Environment Agency

Sunbleak *Leucaspius delineatus*

Description

- Small, slender fish, resembling small bleak *Alburnus alburnus*, but less laterally flattened and with broader head.
- Olive green back, white belly and silvery sides, with iridescent blue sheen along flanks.
- Large head, with protruding lower jaw, upturned mouth and relatively large eyes.
- Distinguished from most other small cyprinids by distinctive, short lateral line, which ends before dorsal fin.

Size

Rarely more than 9cm.

Biology and behaviour

Sunbleak are short lived, becoming sexually mature at the end of the first year of growth and rarely surviving more than four years. Spawning typically takes place between April and August. Before spawning, males defend territories and clean the chosen spawning substrate using their mouths. Females then select a male with which to spawn and he is left to guard the eggs until the embryos hatch. Females develop a swollen genital papilla from which they deposit strips of a maximum of 80 eggs on the stems and leaves of aquatic macrophytes and on flat surfaces of artificial structures such as pier supports. Throughout the spawning period, individual females lay several batches of translucent, highly adhesive eggs. Sunbleak feed mainly on zooplankton and on terrestrial insects taken from the water surface.

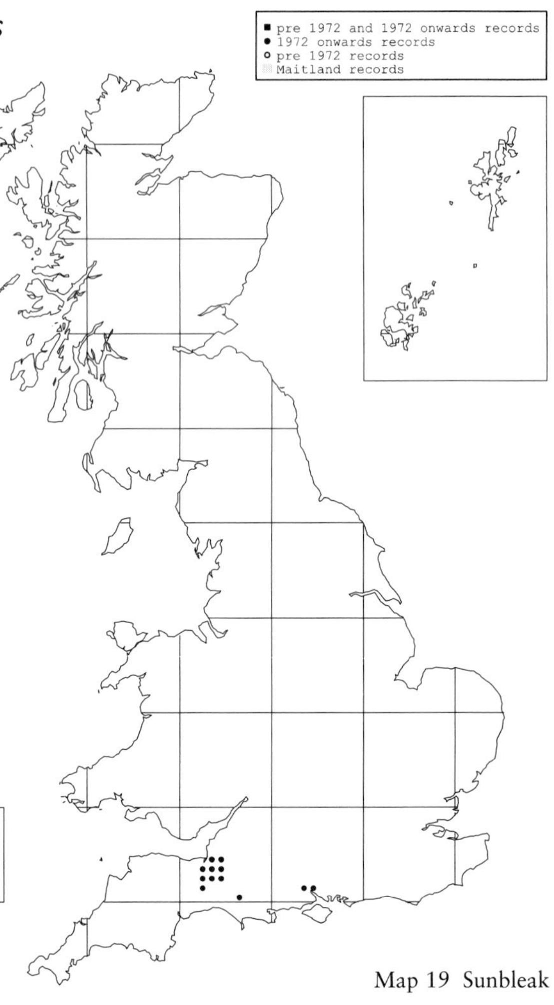

Map 19 Sunbleak

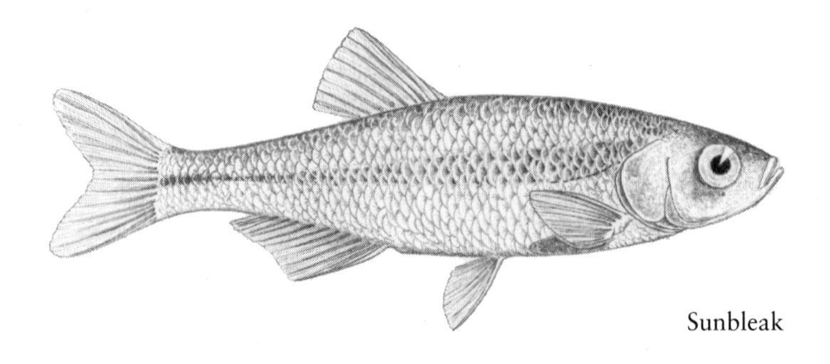

Sunbleak

Habitat

In continental Europe, sunbleak are commonly found in ponds, lakes, drains, canals and lowland rivers, and their associated backwaters. From spring to autumn, large shoals swim in open water close to the surface, but during spawning congregate amongst marginal, aquatic vegetation. During the winter months, when food is scarce, they prefer deeper water. Sunbleak are tolerant of high water temperatures and low levels of dissolved oxygen.

Distribution in Britain

Sunbleak were first introduced to Britain in 1986 by an ornamental fish supplier in Hampshire. From this introduction they dispersed along the neighbouring River Test, and were found in 1987 further down the Test floodplain at Broadlands Lake, where they were originally believed to be bleak. The origin of the population in Somerset is unclear, but they were first noticed there in the King's Sedgemoor Drain in 1990. By 1994, sunbleak were present in the Bridgwater and Taunton Canal and the River Parrett. Sunbleak can now be found throughout the Somerset rivers, drains and connected waterbodies, as well as several still waters in the area. They are also present at several still waters in Hampshire and at least one complex of ponds in Dorset. It is believed they have been spread through fish transfers, where sunbleak have been mistaken for the fry of other cyprinids. Their small size also means that the fish can easily escape from still waters through gratings and screens, thereby becoming established in connecting rivers and their tributaries.

World distribution

Widely distributed across continental Europe, from the Caspian Sea to the North Sea, in several catchments including the Rhine, Volga and Danube. In western Europe sunbleak are indigenous to northern France, Belgium, Holland, Germany, Denmark and southern Sweden. Not naturally present in southern France, or south of the Alps and Pyrenees.

Status

- Listed on the Prohibition of Keeping or Release of Live Fish Orders (see Appendix 3) as a species for which release to the wild is not permitted without a licence.
- Where they are present, sunbleak can be very abundant and often the most numerous species, and they are increasing in range.
- Potential threat to juveniles of native species, because sunbleak of all ages compete for the same dietary items as other juvenile cyprinids.

Hybrids and related species

- No reports of hybrids.
- Sunbleak is the only species in the genus *Leucaspius*. *L. stymphalicus* (Valenciennes) and *L. marathonicus* (Valenciennes), two species thought to be endemic to Greece, have now been reclassified to the genus *Pseudophoxinus* and are no longer considered to be related to sunbleak.

Sunbleak and Man

Sunbleak are of little interest to anglers and, where present, can be regarded as a pest. However, match anglers have been known to target this species in competitions to increase their overall catch. Despite their small size and slight appeal to anglers, sunbleak have several common names, including belica and motherless minnow. The last name refers to its tendency to 'appear' suddenly in large numbers in small pools; it is derived from its common German name, *Moderlieschen*, from the Low German *Moder-loseken* meaning motherless. A further name, sundace, is used in the 2003 Amendment to the Prohibition of Keeping or Release of Life Fish (Specified Species) Order 1998.

Further reading:

Farr-Cox, Leonard & Wheeler, 1996; Gozlan *et al.*, 2003; Pinder, 2001; Pinder & Gozlan, 2003.

Authors: Jonathan Shelley, Environment Agency, & Adrian Pinder, Centre for Ecology and Hydrology

Chub *Leuciscus cephalus*

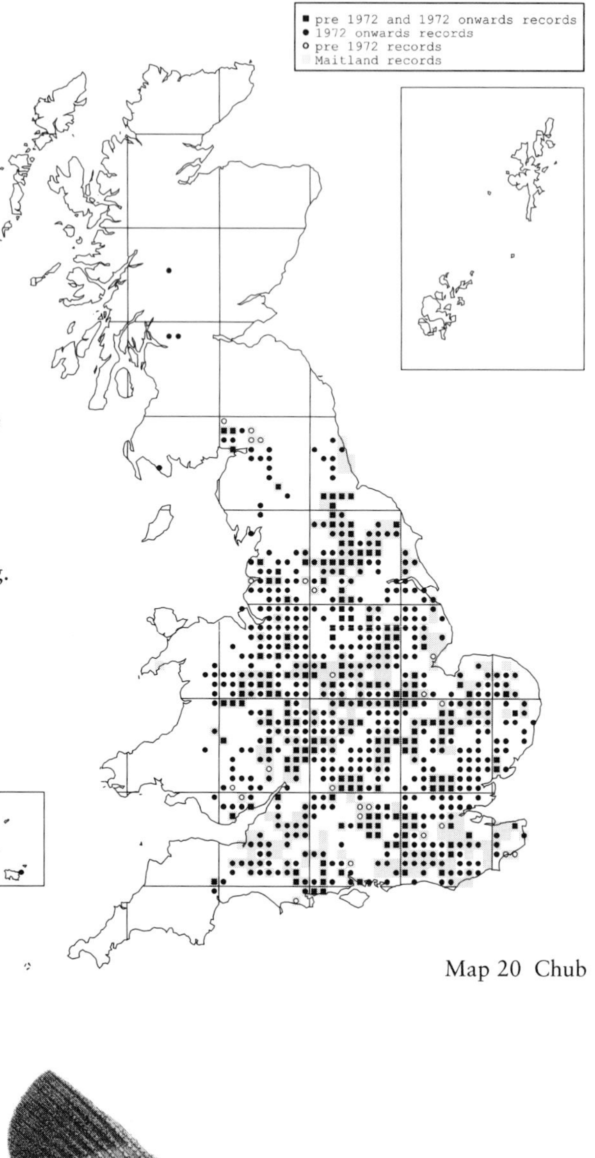

Map 20 Chub

Description

- Body elongate and sturdy, with only slight lateral compression.
- Head blunt and wide across eyes.
- Back greenish-brown, with brassy flanks that fade to off-white underside.
- Fins dark, particularly towards the tips: pectoral and pelvic fins often slightly reddish; anal and dorsal fins have convex rear edge.
- Mouth terminal and large, with clear white lips.
- Scales large, dark edged and distinct.

Size

Normally 30–45cm long, weighing 0.5–1.5kg. Largest British rod-caught specimen weighed 3.9kg.

Biology and behaviour

Males tend to mature at three or four years with the females maturing at four to five years. Spawning usually takes place in shallow riffles over stones, gravel or weed to which the eggs will stick. Chub spawn in May or June, when large shoals congregate around suitable habitat and three or four males can spawn with a single female. The eggs hatch in five to nine days depending on water temperature, and the young fry are swept into slower flowing pools after hatching. Small chub feed on insect larvae and crustaceans and also

Chub

some plant material. Large chub consume much larger prey such as small fishes, frogs and possibly even young water birds, and are reported to feed selectively on crayfish where they are available. Chub will also take fruit and berries when they fall into the water.

Habitat

Chub occur mainly in the lowland and middle reaches of large rivers, where the flow is moderate. They also appear to be attracted to small tributaries; surprisingly large individuals can be found in small channels of the main river. They shoal readily when small and for spawning, but larger individuals are often solitary, tending to be found in deeper pools especially where there is some cover.

Although normally associated with flowing waters, chub appear regularly in anglers' catches from some still waters. Whether chub can breed successfully in still waters remains uncertain, but good growth rates have been recorded.

Distribution in Britain

Chub are found throughout much of southern Britain, but are absent from the far south-west of England, the westerly flowing rivers in Wales and from most of Scotland. The overall distribution in Britain has changed little since 1972, but chub are increasingly recorded in still waters where they have been introduced for angling.

World distribution

Chub are one of the most widespread river fishes in Europe, occurring naturally from southern Sweden to Turkey and from Portugal to the Urals, but they are absent from Ireland and the Mediterranean islands. They are also found in some of the large European lakes, such as Lake Garda in northern Italy (where they can be seen feeding on titbits from restaurants on the lake shore), and there are records from brackish water in the Baltic Sea.

Status

- Not threatened in Britain, the rest of Europe or globally.

Hybrids and related species

- Hybrids with several other cyprinids have been recorded, including roach, rudd and bleak.
- Closely related to orfe and dace, both of which are widespread in continental Europe, and to several other *Leuciscus* species with highly localized distributions, mainly in the Balkans.

Chub and Man

Chub are popular sport fishes wherever they are found, and are farmed commercially for restocking fisheries. With soft, mushy flesh full of sharp bones, chub are generally considered not worth eating, despite Izaak Walton's claims to the contrary with two detailed recipes.

Author: Ian Wellby, Brooksby Melton College

Orfe *Leuciscus idus*

also known as Ide

Description

- Body thickset, with some lateral compression.
- Head comparatively large in relation to body, with noticeable hump behind head.
- Mouth broad and terminal.
- Dorsal fin has slightly convex or straight edge; anal fin has very slightly concave edge.
- Wild form has greyish back which fades to silver sides and white underside.
- Several colour varieties including gold and blue.

Size

Varies considerably with habitat. Wild specimens normally grow to 40cm in length and 1.2kg in weight.

Biology and behaviour

Spawning takes place when water temperature rises above 10°C, usually in April or May. Like dace (see p. 79), male orfe become very rough to the touch as spawning tubercles develop on the head and body. The males migrate to spawning areas, usually somewhere with a stony substrate, several days before the females. As soon as the females arrive, the males drive them to a suitable location for spawning where the water is well oxygenated, for example in shallow water near a lake shore or in fast-flowing water. The number of eggs laid depends on the size of the fish. Typically between 39,000 and 115,000 yellow, round, sticky eggs are laid, which fasten to rocks and weeds and hatch in 15 to 20 days. Rates of growth vary with habitat and food availability, but adults normally mature in five to seven years. Orfe are omnivorous, usually feeding and moving in shoals, consuming most food items, from algae to mayfly larvae and including their own eggs. Large, often solitary specimens eat other fishes, mainly fry and smaller species.

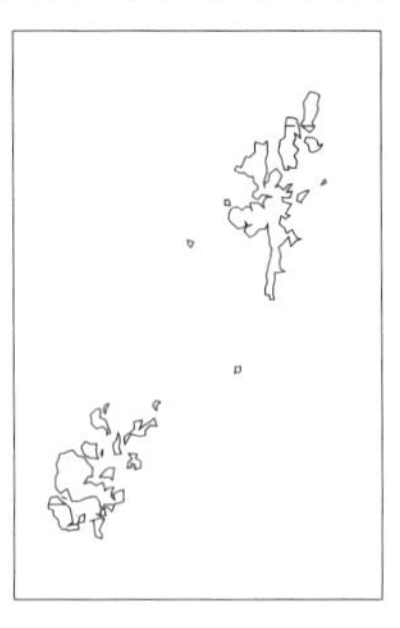

Map 21 Orfe or Ide

Habitat

Orfe prefer clean, slow-moving middle and lower reaches of rivers and lakes, but can survive almost anywhere, including garden ponds and canals. They prefer cooler water, from 4°–20°C, but appear to acclimate to higher temperatures. In their natural range they migrate upstream to headwaters to spawn. Orfe have been introduced successfully into a wide range of habitats to form self-sustaining populations, including in some lakes and rivers in Britain.

Distribution in Britain

The first successful introduction of orfe to Britain was probably in 1874, when over 100 juveniles were brought from Wiesbaden in Germany to Woburn Abbey, Bedfordshire. Feral populations now occur at scattered sites, mainly in England, and the species is being increasingly recorded in many areas of England and Wales as it is introduced for both sporting and ornamental purposes. However, there are only a few records from Scotland.

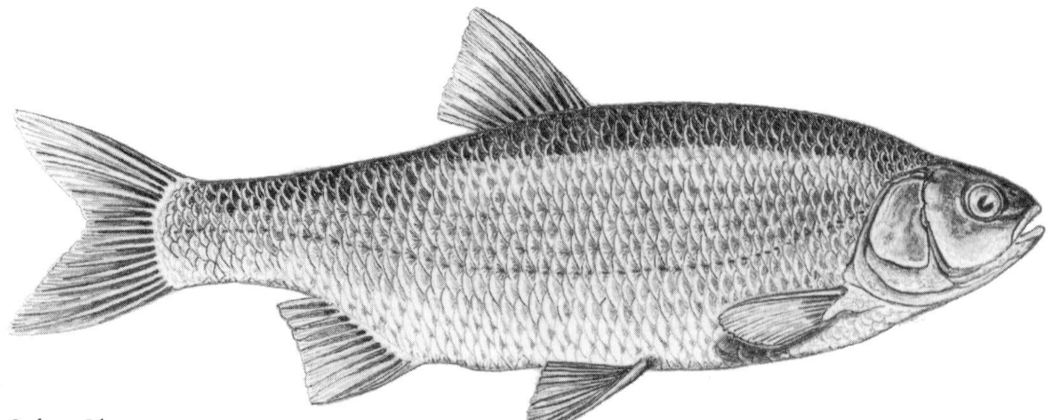

Orfe or Ide

World distribution

Orfe occur naturally in most of Europe east of the Rhine, in Sweden and Finland, and in Asia south to the northern shores of the Black and Caspian Seas. They can also be found in the brackish waters of the Baltic Sea. Orfe are absent from Ireland, Norway, Switzerland, the southern Alps and the rest of southern Europe. However, as in Britain, they have been introduced to and may be established in several parts of northern France, the lower Loire valley and even in Corsica.

Status

- Introduced and not threatened in Britain.
- Not threatened in the rest of Europe or globally.

Hybrids and related species

- No hybrids have been recorded in Britain but, in the right environment, orfe may hybridize with other cyprinid species to produce infertile offspring.
- Closely related to chub and dace.

Orfe and Man

Orfe are fished for sport wherever they are found. As they are active, fighting fishes and help to make up good match weights in still waters, their popularity is increasing at present in Britain. Orfe are fast growing and hardy and able to digest food at low temperatures, which makes them popular with fisheries' managers. A specimen orfe can grow up to 8kg, although the British record is just under 4kg. They are also farmed extensively for the ornamental trade, particularly the golden and blue colour forms. Their attraction as ornamentals has increased because they become used to humans and can be hand fed. They are edible and are farmed and harvested for the table, particularly in Finland and Russia.

Authors: Ian Wellby & James Vickers, Brooksby Melton College

Dace *Leuciscus leuciscus*

Description

- Body slender and compressed laterally.
- Head small and pointed, with inferior mouth; eye with distinct yellow iris.
- Green/brown back, striking silver flanks and white underbelly.
- Pectoral, pelvic and anal fins pale, occasionally with yellow or orange tint; hind edges of dorsal and anal fins concave; anal fin starts behind dorsal fin.

Size

Normally up to 20–25cm long; weight up to 600g.

Biology and behaviour

Dace are usually the first of the cyprinid species to spawn, normally between February and early May when water temperatures range between 8° and 14°C. During this period males, which mature at about two years, develop spawning tubercles on the head and flanks. Dace spawn communally in gravelly, well-oxygenated shallows of rivers and streams. Females, which

Dace

are normally sexually mature at three years, deposit between 2,000 and 27,000 golden to salmon-pink, sticky eggs. The fertilized eggs hatch after about 25 days and the growing larvae gather in large shoals, feeding on tiny zooplankton. Adult dace eat a variety of insects and crustaceans, graze on algae and water plants and will take terrestrial items such as flies, earthworms and berries that fall into the water. Unlike most British cyprinids, dace continue to feed opportunistically, although at a reduced rate, throughout the winter.

Habitat

Dace occur naturally in clean, fast-flowing rivers and streams almost throughout their length, but particularly in the middle and lowland reaches. They form large, active shoals, usually in the upper layers of the water, but larger specimens are normally more solitary. Dace appear to do well in still waters where they have been introduced, although it is uncertain whether they breed successfully. Large specimens are known from some still waters, such as Llandegfedd Reservoir, near Pontypool in South Wales.

Distribution in Britain

Dace are now found throughout the rivers of England and eastern Wales, including the Dee and its tributaries. The use of dace as live bait by anglers has resulted in the species being introduced into much of its present range in Britain, including still waters. In Scotland, the occurrence of dace in the Rivers Clyde and Endrick, and a few rivers in the south, all result from introductions.

World distribution

Dace are found in river systems throughout most of Europe with the exception of Portugal, Spain, Italy, the countries bordering the Adriatic, and much of Greece. In many of these areas where dace are absent, there are several other *Leuciscus* species, some of which are very localized. Dace are also absent from the west and extreme north of Scandinavia. Dace continue to occur naturally eastwards into the river systems of Siberia flowing to the Arctic Ocean. As the result of an accidental introduction in 1889, dace occur in the River Blackwater in southern Ireland.

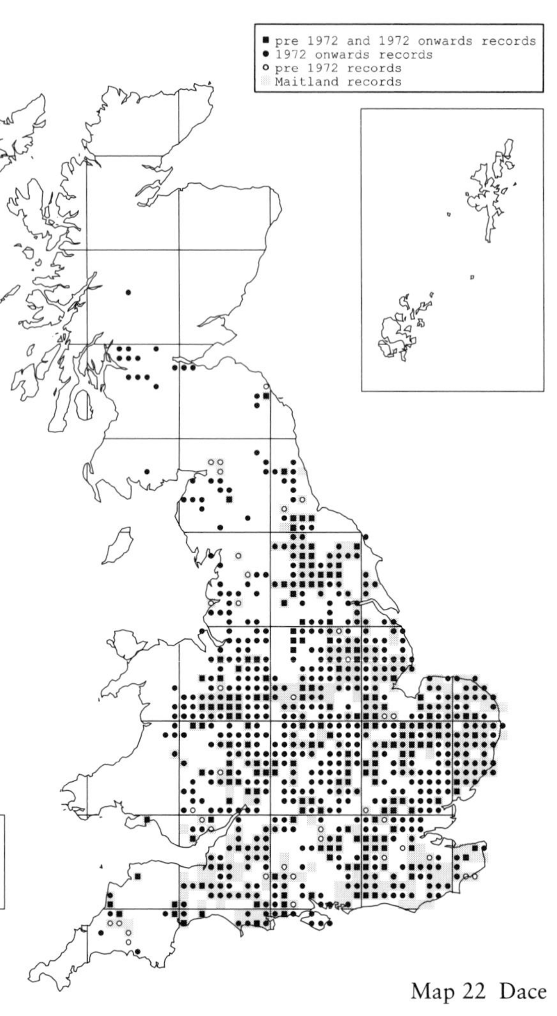

Map 22 Dace

Status

- Not threatened in Britain, the rest of Europe or globally.

Hybrids and related species

- Hybrids have been recorded with roach and it is possible that hybridization occurs with other cyprinid species.
- Closely related to chub and orfe.

Dace and Man

The use of dace by anglers as live bait for predatory species such as pike has greatly extended their distribution, but other than this the species is not classed as either a good sport fish or good eating, due to its small size. Dace are occasionally regarded as a pest in some countries, but there is some commercial exploitation in parts of Russia. In Britain, dace are farmed for restocking rivers that have been affected by pollution, but this is not done on a large-scale basis.

Authors: Ian Wellby & Ian Cook, Brooksby Melton College

Minnow *Phoxinus phoxinus*

Description

- Small and slender with rounded cross-section and blunt snout.
- Fins short-based and rounded.
- Dark, mottled back, often with distinctive black and gold stripe along middle of flank. Males develop red underside at spawning time.
- Very small, virtually invisible scales.

Size

Maximum length up to 12cm but rarely more than 6–9cm.

Biology and behaviour

Minnows are small and short-lived, rarely living beyond three years. They mature in their first year and spawn in late spring, once water temperatures reach 14°–16°C, in shallows with well-oxygenated water and a gravel or small pebble substrate. In lakes, spawning can be observed around the mouths of inflowing streams. A female minnow of around 7cm in length will deposit around 1000 eggs, each 1.5mm diameter, amongst gravel. Minnows live in large shoals, spending most of their time in the margins feeding on a wide variety of small invertebrates, algae and plant debris, and will readily take items from surface, water column or substrate. Whilst minnows are most active in the warmer months, they can feed at temperatures as low as 4°C.

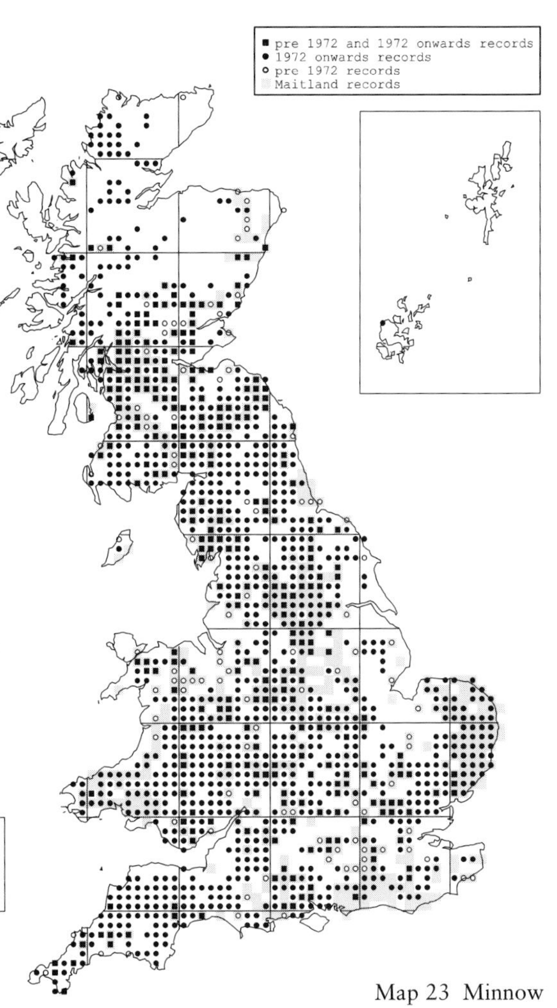

Map 23 Minnow

Habitat

Minnows prefer clean, well-oxygenated water with a gravel substrate. Typically they inhabit the upper and middle reaches

Minnow

of rivers and their tributary streams, but they also occur in clean, gravel- and stony-bottomed lakes. Provided oxygen levels are high, they are able to tolerate a certain amount of organic pollution; in semi-polluted rivers, they are particularly abundant downstream of weirs where they benefit from the abundance of small food items.

Distribution in Britain

The minnow is now very widely distributed throughout Britain, its natural range having been extended due to introductions many years ago to westerly-flowing river systems. Currently it is absent from large areas of north-east Scotland and occurs only sporadically in the Fens in eastern England. Since 1972, there appears to have been a contraction in the range of the species in parts of Sussex, Kent, Norfolk, and the lower Trent catchment. The reason for this is unclear, but changes in land-use may have destroyed the favoured gravel spawning substrate in small rivers, though it may be simply a reflection of insufficient recording in some areas.

World distribution

Widespread and ubiquitous in Ireland and much of continental Europe through to Russia, but not in western and northern Scandinavia, Italy, the Iberian Peninsula, the southern tip of the Balkan Peninsula or Iceland.

Status

- Not considered rare or threatened in Britain or the rest of Europe.

Hybrids and related species

- The minnow has not been recorded as hybridizing with any other species.
- Two other species in the genus *Phoxinus*, Czekanowski's (or Poznan) minnow *P. czekanowskii* Berg, and the lake or swamp minnow *P. perenurus* (Pallas), which are confined to north-eastern European rivers draining into Baltic Sea or the Arctic Ocean. The latter is sometimes misspelt *percnurus* and some authors now put this species in the genus *Eupallasella*.

Minnows and Man

The minnow is of no commercial and limited culinary value and is not considered as a sport fish by recreational anglers; indeed, it is often regarded as a nuisance by those seeking other species. However, it is a very important food species for larger fishes such as perch, trout, pike and chub. Being plentiful and visible in the margins of cleaner rivers, minnows are frequently caught by young children with small nets and so can be considered to have a significant amenity value.

Author: Graeme Peirson, Environment Agency

Topmouth gudgeon
Pseudorasbora parva

Description

- Small cyprinid with upturned lower jaw.
- Relatively large scales with dark edges.
- Back brown, belly white, sides silvery with iridescent violet sheen; more subtle colours (yellows and pinks) during spawning.
- Band of dark (sometimes purple) pigment extends along mid body from eyes to base of tail; more obvious in smaller individuals and often absent in fishes more than 6cm total length.

Size

Length up to 10cm.

Biology and behaviour

A short-lived species, rarely surviving for more than three years, topmouth gudgeon become sexually mature at the end of the first year's growth. They spawn between May and mid-summer, with individual females laying several batches of sticky eggs, in short strips, throughout this period. Males are territorial during spawning and develop secondary sexual characters, such as dark coloration and tubercles around the mouth and below the eyes. Males entice several females to spawn on the substrate of the territory, which can include stones, macrophytes, floating objects and artificial structures. The male then guards the eggs until they hatch, taking approximately seven days at 20°C. Topmouth gudgeon feed on a wide range of items including algae, benthic invertebrates, zooplankton, molluscs, and the eggs and larvae of other fishes. As opportunistic feeders, they will also feed on terrestrial insects and other floating objects from the water surface: this is accompanied by an audible clicking noise (hence an alternative name 'clicker barb').

Habitat

Topmouth gudgeon occur in a wide range of habitats, including rivers, lakes, canals, drains and ditches, although it is reported to favour calm, slow-flowing waters.

Distribution in Britain

Topmouth gudgeon were introduced to Britain in the mid- to late 1980s by a supplier of ornamental fishes. This alien species is now well established at the original location near Romsey, Hampshire, as well as at a still-water site near High Wycombe, Buckinghamshire, and several locations in Staffordshire. More recently, further populations have been discovered at still waters in Cheshire and Cumbria. Its small size makes it difficult to distinguish from other small cyprinids and the initial dispersal within Britain was probably by unintentional transfers with other species of fish.

World distribution

Asiatic in origin (Japan, China, Korea and the Amur basin), topmouth gudgeon were accidentally introduced to Romania in 1960, mixed with a consignment of fry believed to contain commercial Chinese carp species. It has since spread throughout the Danube basin and can now be found in most of Europe and parts of North Africa. This rapid dispersal is believed to be a consequence of both natural expansion and unintentional transfers.

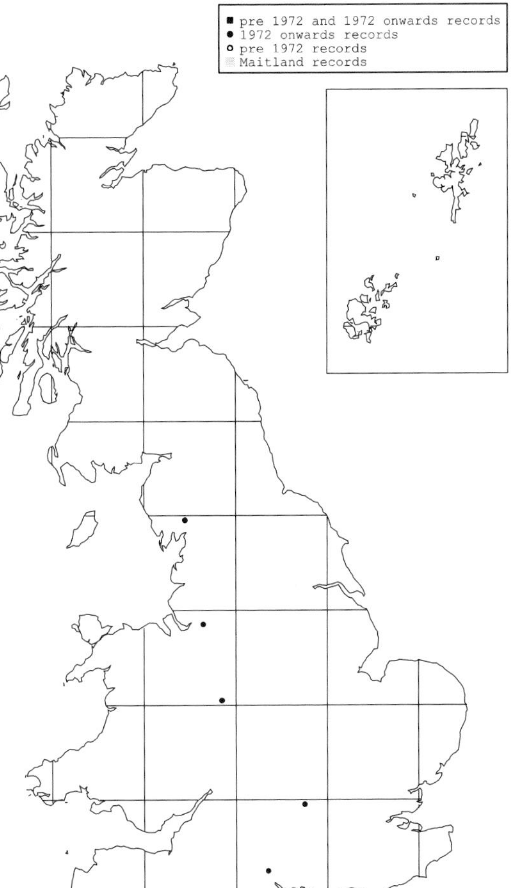

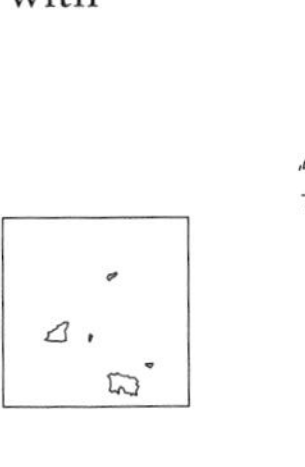

Map 24 Topmouth gudgeon

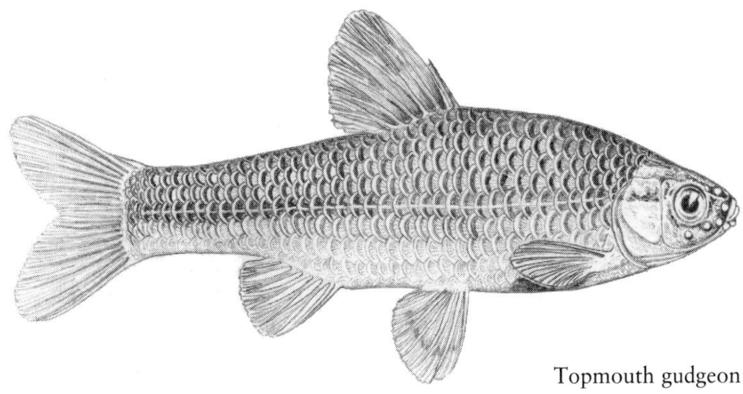

Topmouth gudgeon

Status

- Regarded internationally as a pest, although risks and impacts need to be quantified.
- Listed under the Wildlife and Countryside Act 1981 and on the Prohibition of Keeping or Release of Live Fish Orders (see Appendix 3) as a species for which release to the wild is not permitted without a licence.
- Highly invasive because of its small size, broad range of environmental tolerance limits and rate and mode of reproduction.
- Potential threat to native species, by direct predation on eggs and larvae, and indirect competition for prey items.
- In Europe, topmouth gudgeon have been shown to be a vector for parasites such as the swim-bladder nematode *Anguillicola crassus* Kuwahara, Niimi & Itagaki and the fluke *Clinostomum complanatum* Rudolphi, and a carrier of pike fry rhabdovirus (PFRV).

Hybrids and related species

- No reports of hybrids.
- Not related to similarly named species (for example *Rasbora* species or gudgeon *Gobio gobio*).

Topmouth gudgeon and Man

The species is popular with aquarists (among whom it is also known as the stone moroko, false harlequin, Japanese minnow, sharpnose gudgeon and clicker barb). This popularity probably led to the original introduction and may have contributed to dispersal of this species in Britain and elsewhere.

Further reading:

Gozlan *et al.*, 2002; Katano & Maekawa, 1997; Maekawa *et al.*, 1996; Pinder & Gozlan, 2003; Rosecchi *et al.*, 1993; Rosecchi *et al.*, 2001; Xie *et al.*, 2001.

Author: Adrian Pinder, Centre for Ecology and Hydrology

Bitterling *Rhodeus sericeus*

Description

- Small, deep bodied, laterally compressed, with large scales.
- Body colour varies with sex and time of year. Overall pink to light purple with distinctive blue-green line along mid-line from centre of the body to tail.
- Mature females have protruding, pale ovipositor at the genital opening, even outside the breeding season; however from April to June the ovipositor extends up to 15–20mm.

Size

Up to 7cm in length, but generally less than 5cm in British rivers.

Biology and behaviour

Bitterling have a unique symbiotic reproductive strategy. In early spring, males defend territories around freshwater mussels. Responsive females are led to the mussel, where they deposit eggs from an extended ovipositor, which is inserted into the mussel's exhalant siphon. The large, oval eggs are lodged either singly, or a few at a time, into the gill cavity of the mussel after which the male bitterling ejects his sperm into the mussel's inhalant water current, so that fertilization takes place within the gills of the host. Larval bitterling develop safely within the mussel for up to six weeks, emerging once the yolk sac has been absorbed fully, by which time they have reached a length of 10mm. After mating, the adult bitterling play an important role in dispersing the mussel larvae which have attached themselves to their gills. Bitterling are omnivorous, but they have an exceptionally long gut which aids the digestion of filamentous algae, such as blanket weed *Cladophora* spp., a commonly used food source.

Bitterling

Habitat

Bitterling are dependent entirely on sites that contain populations of freshwater mussels (see above). They are found in lakes, slow-flowing lowland rivers, fenland drains and canals. They can survive in waters that suffer from periodic drops in oxygen and which would prove fatal to most other species. This is due to their unusual metabolism, shared with goldfish and crucian carp, which enables them to respire in conditions with very low levels of dissolve oxygen without polluting their own body tissues. This unusual ability also enables bitterling in their first year to overwinter buried in mud.

Young bitterling form large shoals with other coarse fishes, including roach and perch. During the spring and summer, adult bitterling move close to the river- or lake-bed where courtship and spawning takes place.

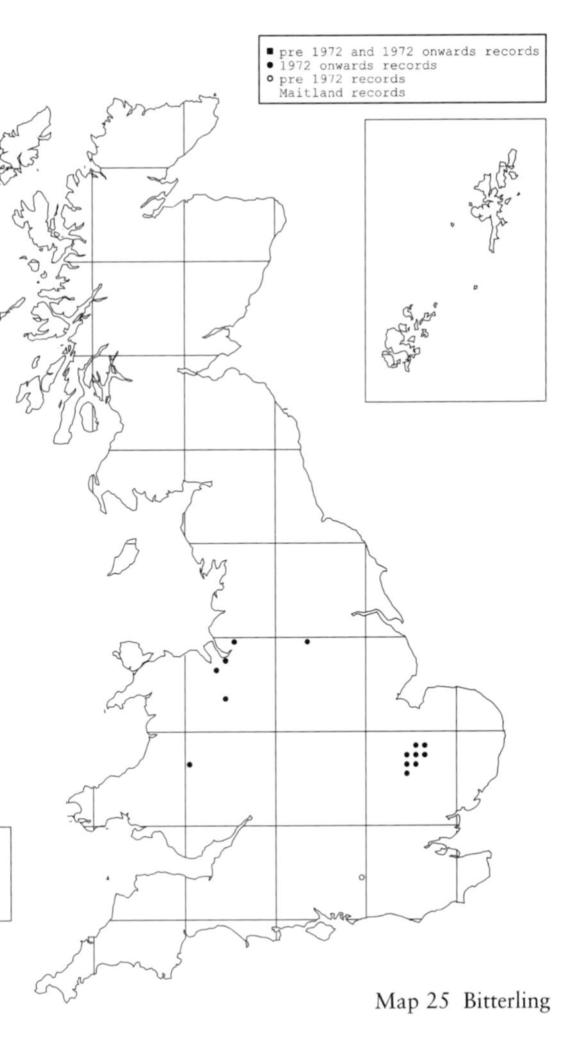

Map 25 Bitterling

Distribution in Britain

The first published British record, in 1954, includes a long history of references to 'strange types of fish' in still waters in Lancashire. It is thought that the first reliable records for Lancashire and Cheshire are from the 1920s and '30s, and that the bitterling population in these counties has persisted through to the present. Further introductions, including some to the Cumbrian lakes, have been reported as being unsuccessful. It is only the Lancashire/Cheshire population, and a second population in Cambridgeshire that are believed to have established themselves in great numbers. The Cambridgeshire population is thought to have been introduced into the Great Ouse system in the mid- to late-1970s and in places bitterling are numerically the second most abundant fishes after roach. Bitterling are continuing to spread through the Great Ouse and associated water bodies. Bitterling populations have become established in smaller numbers at other sites throughout Britain. In particular, there are numerous reports from large ornamental ponds, but these have not been recorded properly.

World distribution

It is thought that bitterling originated in eastern Asia, where up to 60 species belonging to three related genera occur. Of these species, the European bitterling *R. sericeus* is the most widespread. Its natural distribution stretches eastwards from north-east France, through Europe, to the basins of the Black and Caspian Seas. It is also native to central and north-east Asia, but absent from Scandinavia. The worldwide distribution of *R. sericeus* has been further increased through introductions. It was first reported in the United States in 1925 and persists in New York State, but only in small numbers.

Status

- Introduced to Britain and currently spreading. Locally abundant, but not believed to be a threat to native fauna and flora.
- Listed under the Wildlife and Countryside Act 1981 and on the Prohibition of Keeping or Release of Live Fish Orders (see Appendix 3) as a species for which release to the wild is not permitted without a licence.
- Although it is protected under Habitats and Species Directive and the Bern Convention, these do not apply to Britain because the species is not native here.

Hybrids and related species

- Not known to hybridize with any other European species.

Bitterling and Man

The attractive coloration and behaviour of bitterling make them a popular aquarium fish. Indeed, it is probably through the aquarium trade that bitterling were introduced into Britain. In the early twentieth century, bitterling were occasionally used as a human pregnancy test; the ovipositors of female bitterling would elongate in the presence of human urine containing oestrogenic hormones.

Further reading:

Aldridge, 1997, 1999; Hardy, 1954; Myers, 1925; van Waade *et al.*, 1993; Wiepkema, 1961.

Author: David Aldridge, University of Cambridge

Roach *Rutilus rutilus*

Description

- Deep body with forward facing mouth.
- Red eyes, dark back with a blue or green iridescent sheen, bright silver flanks, dark dorsal and caudal fins with reddish brown pectoral, pelvic and anal fins.
- Single dorsal fin above the base of the pelvic fins.

Size

Adults range from 20–40cm in length and 1–2kg in weight.

Biology and behaviour

Roach spawn once a year during spring when water temperatures rise to approximately 14°C. Gravid roach locate shallow areas with dense submerged vegetation before shedding eggs indiscriminately over the weed: the eggs are 'sticky' and readily adhere to vegetation. A large female roach can produce over 200,000 eggs. In fast-flowing rivers, roach will often choose to spawn on willow moss (*Fontinalis* spp.), which is typically found growing on man-made structures including weir sills, undershot hatches and the footings of road bridges. Young roach will shoal in large numbers and have a limited home range early in life. These juveniles feed on zooplankton and other small invertebrates, while older fishes prefer larger benthic invertebrates and filamentous

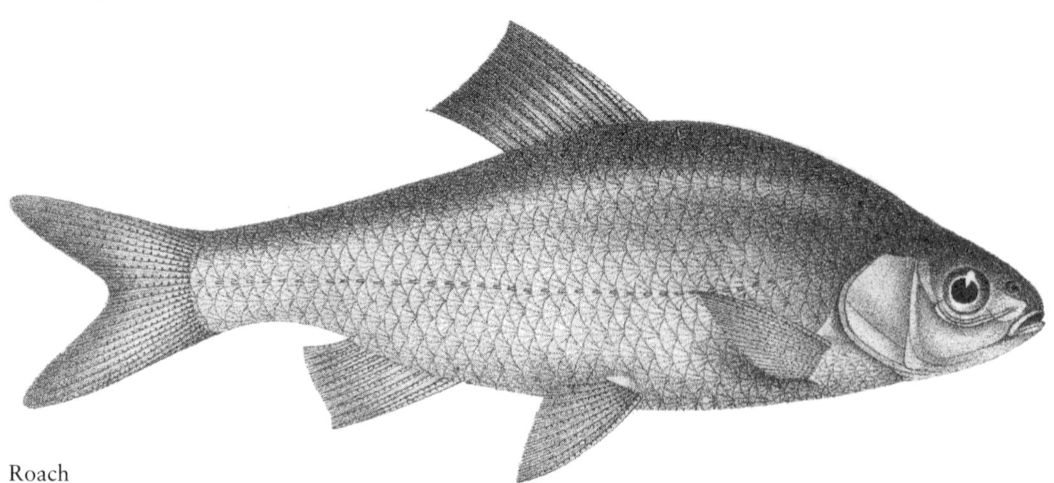

Roach

algae. Large adults tend to live in small groups away from areas frequented by juveniles. Roach are not territorial, but will migrate to preferred habitat types on both a daily and a seasonal basis.

In favourable habitats in southern Britain, roach can live for up to 18 years but in other areas, more typically, they live for up to 14 years. Male roach can mature in their second year, but maturity usually occurs after three years, and in females after four years. Roach can attain one kilogram in weight within eight years, but the availability and quality of food in different habitats and population densities influence growth rates.

Habitat

This prolific species generally inhabits lowland rivers and still waters, and is particularly tolerant of poor water quality in intensively managed catchments. As a result, roach are found in a wide variety of waters ranging from pristine chalk streams to eutrophic village duck-ponds. They are also abundant in brackish water, including inter-tidal reaches of many rivers, such as the lower Thames, Dorset Stour and Bristol Avon.

Juvenile roach are most commonly associated with shallow marginal areas, preferring dense weed cover and low water velocities. These conditions provide both cover from predators and a ready supply of food. Roach tend to be less numerous in clear water environments; this may reflect adverse conditions for the survival of juveniles. Adults are found in both shallow and deep water, including still, slow-moving or fast-flowing waters. The very largest fish seek deep, undisturbed areas in still water and less turbulent refuges in rivers.

Distribution in Britain

Roach are widespread throughout Britain being most concentrated in the north, south and east of England, and less so in the south-west. Roach are also present in many waters throughout Wales and lowland Scotland. The distribution of roach is a result of their natural preference for lowland habitats and their introduction to waters that had not previously contained them. Their adaptable nature enables them to be successful in most waters to which they have been introduced.

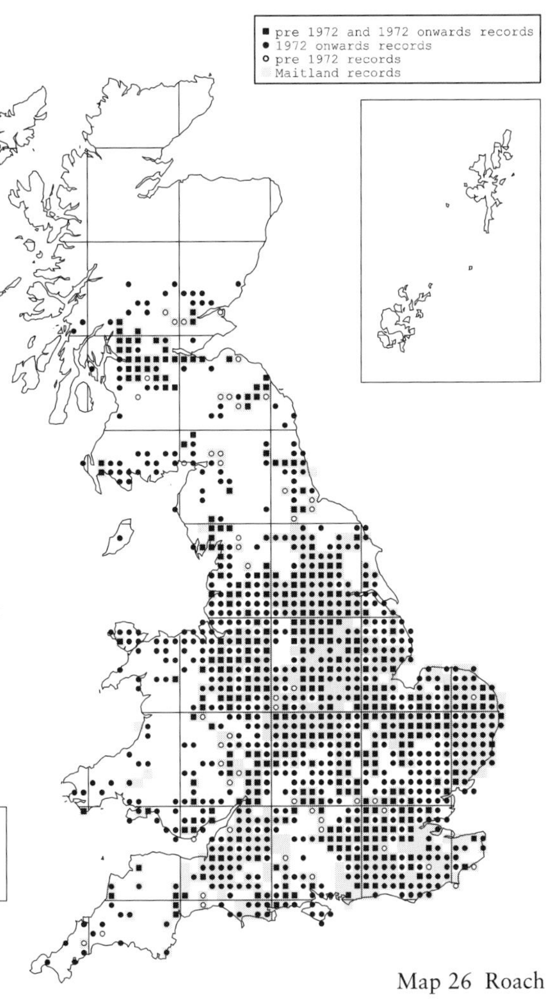

Map 26 Roach

World distribution

Roach are principally distributed throughout Europe eastwards into Asia, and from the Baltic Sea, Sweden and Finland in the north to European Turkey in the south. They have been introduced into the Iberian Peninsula. Roach are not native to Ireland but are now widespread, having first been noted during the 1960s in the Blackwater system in the south-west and then in central and northern Ireland during the 1970s, after a series of illegal introductions by pike fishermen.

Status

- Not threatened in Britain or continental Europe.
- One of the most gregarious freshwater fish species in Britain.

Hybrids and related species

- The only British species of roach although there are up to seventeen other species in the genus *Rutilus* in continental Europe.
- Hybrids are common, especially roach × bream and roach × rudd.
- Less common hybrids include roach × chub, roach × bleak and roach × silver bream.

Roach and Man

In earlier times, large quantities of juvenile roach were netted for food. Whilst this is no longer the case, the adult roach has become a very popular sport fish for anglers, being returned to the water after capture. As a result roach are an important economic component of the freshwater angling industry that serves nearly four million people in Britain.

Further reading:

Mann, 1973.

Author: Matt Carter, Environment Agency

Rudd
Scardinius erythrophthalmus

Description

- Fairly deep bodied and laterally compressed, with upturned mouth.
- Head and back dark greenish brown, grading through silvery gold on flanks, to white on belly. Iris deep gold with red spot on upper side.
- Fins bright red, especially ventrally .
- Base of dorsal fin set markedly behind line of base of pelvic fins.
- Well-developed, scale-covered keel between pelvic fins and anus.

Size

Normally up to about 20cm long and weighing about 200–400g; exceptionally more than 35cm and over 1kg in weight.

Biology and behaviour

Both sexes mature at two to four years old, although males often mature a year earlier than females. Spawning commences once temperatures reach 15°–16°C, normally between April and July, when the female may produce from 90,000 to over 200,000 eggs, depending on weight. Spawning tends to be earlier in the south and east than in the north and west. Rudd shoal in shallow areas, where the sticky eggs adhere to submerged weeds. On hatching, the fry, which

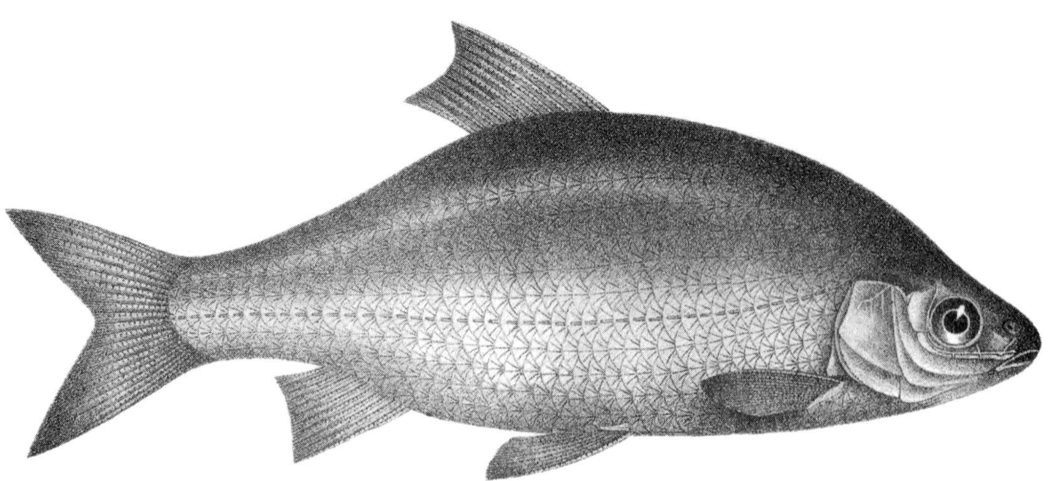

Rudd

are 4–6mm long, remain attached to the weed until the yolk sac is fully absorbed before starting to swim and feed. Rates of growth of rudd are variable, depending on local conditions; if these are favourable, they may reach up to 20cm in three years. Populations consisting only of stunted fishes less than 10–12cm can occur, often in small areas of water with a high population density.

Rudd feed mainly near the surface and the middle of the water column. Young rudd take algae, protozoa, rotifers and small crustaceans, primarily from the plankton. As they grow, bottom- and weed-dwelling invertebrates become an important element in their diet. Terrestrial insects taken from the water surface are also important prey. Small quantities of filamentous algae and higher plants may also be consumed.

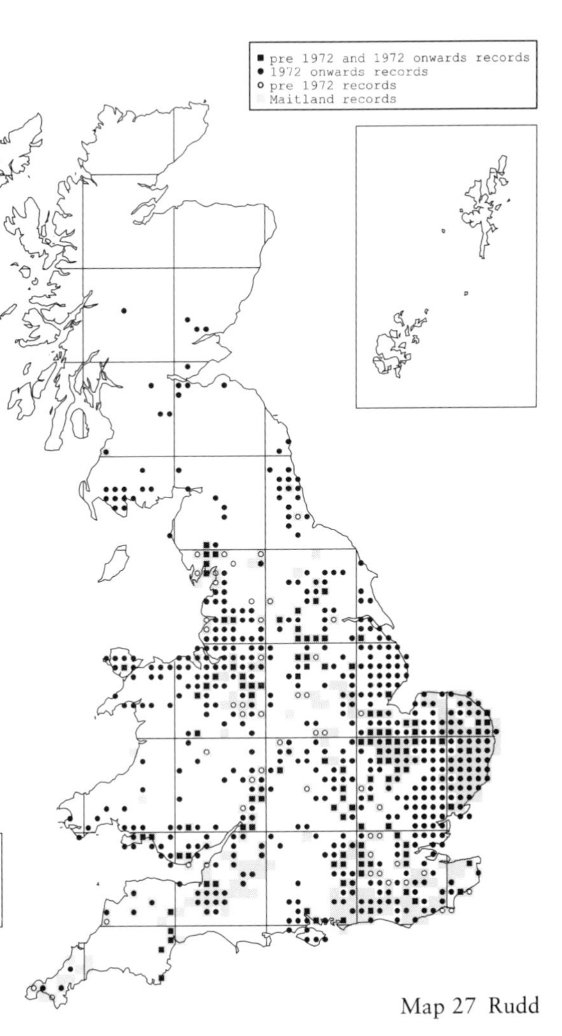

Map 27 Rudd

Habitat

Rudd inhabit slow-flowing rivers, lakes and ponds where there is abundant submerged vegetation. They are tolerant of a wide range of conditions, including brackish water and small ponds, but prefer clear water and have been shown to decline in numbers if the water becomes clouded by turbidity or algal growth resulting from silty run-off or eutrophication.

Distribution in Britain

Rudd are probably native only to south-east England, but they have been redistributed very widely by and for anglers. They now occur in lowland areas of the Midlands, north-west and north-east England, Wales and south-west Scotland, with notable extensions of range even since 1972 in these areas and in Lincolnshire. The absence of rudd from much of mid-Wales and the upland areas of England is partly due to a lack of suitable habitat, particularly of slow-flowing rivers and shallow lowland lakes. Their continued spread into Scotland, where they have appeared only recently, may be similarly limited.

World distribution

Rudd occur throughout much of Europe, from Ireland eastwards to the Aral Sea. They are absent from Spain and Portugal, northern Russia and much of Scandinavia, except Denmark and the southern parts of Sweden and Finland.

Status

Threatened throughout its range by habitat deterioration or loss caused by eutrophication, especially of large shallow lakes.

Hybrids and related species

- Commonly hybridizes with other cyprinid species, principally roach and bream. Hybrids with bleak, dace and silver bream are also known.
- Closely related to other cyprinids, including one or more European species of *Scardinius* that do not occur in Britain.

Rudd and Man

Rudd are a very popular sport fish for anglers, both as a quarry in their own right and as bait fish for pike. They are not palatable and have not been used for food in Britain. They are easy to keep in captivity and, with their bright coloration and tolerance of conditions in aquaria and small ponds, have gained some favour as an ornamental species. Selective breeding has led to the development of strongly coloured varieties, often known as golden rudd.

Author: Robin Burrough, Environment Agency

Tench *Tinca tinca*

Description

- Body thickset, laterally compressed, giving heavy appearance. Mouth small, thick lipped, with well-developed barbel at each corner. Eyes small, orange/red.
- Olive green, but can vary from golden to inky black/green.
- Dorsal fin high, arched. Caudal peduncle broad. After sexual maturity at 3–4 years, males can be distinguished by their long pelvic fins, with a thickened front ray, which extend almost to the vent.
- Scales small, elongate oval, with thick protective mucus.

Size

Maximum length about 70cm. Tench are sexually dimorphic, females being larger; in exceptional circumstances growing to over 7kg, although a more typical weight would be less than 3kg.

Biology and behaviour

Females may carry 80,000 to 750,000 eggs. Spawning, on aquatic plants in shallow water, takes place between May and July and may be protracted. Tench require higher temperatures to spawn than other cyprinids, typically 18°–24°C. The temperature requirements and later spawning often result in a shorter growth period than other cyprinids. Growth rates, linked to temperature in early juvenile stages, are relatively slow, typically 3–4cm, 7–8cm and 12–15cm in the first, second and third year of growth respectively, but tench are long-lived and can live up to 20 years.

Tench feed mainly near the bottom, in the substrate or at the base of the water column, amongst aquatic plants, grazing from the plant surfaces and in the substrate on aquatic insect larvae, zooplankton, algae and molluscs. Tench have four or five pharyngeal teeth, which break up the food prior to being ingested. They generally tolerate water temperatures between 4°–25°C. Below 4°C, tench become torpid and may remain, without feeding, on the bottom or within a muddy substrate throughout cold weather.

Habitat

Tench are primarily found in slow-flowing lowland rivers, still waters and canals, preferably with good communities of macrophytes and a muddy substrate. Habitat complexity is important for their feeding strategy (grazing), the presence of prey, and avoidance of predators. They are tolerant of low levels of dissolved oxygen and consequently are often the main survivors of pollution incidents.

Distribution in Britain

Tench are found throughout most of southern Britain, but are absent from upland areas of Wales, much of Scotland, and all offshore islands. Areas where tench are found commonly – in Somerset, the Midlands, East Anglia, Lincolnshire and Lancashire – reflect its preference for lowland rivers, canals and still waters. They are absent from most upland areas because the habitat is unsuitable, and have failed to colonize isolated waterbodies. Comparison with the distribution mapped in 1972 shows a broadly similar range now, but with limited expansion in the north and in Wales. There appears to have been some contraction in range in the south and south-west of England, possibly because some earlier introductions were not viable.

World distribution

Tench are widely distributed throughout Eurasia, though absent from most of Scotland, northern Scandinavia, the west Balkans and most of Turkey. They are also widely distributed throughout North America, through extensive introductions from Germany by the US Fish Commission in the late nineteenth century.

Status

- Not threatened or protected in Britain, the rest of Europe or globally.

Hybrids and related species

- Does not hybridize with other cyprinids.
- Colour variations exist, including golden tench.

Tench and Man

In the past it has been known as the 'nurse fish' or 'doctor fish'; names attributable to the purported healing properties of the mucus on the skin surface. Tench are often stocked together with carp in ponds and other still waters in many parts of continental Europe, where they are eaten in a variety of ways including baked and pan-fried. The golden variety of tench is stocked as an ornamental fish in ponds throughout Europe.

Further reading:

Kennedy & Fitzmaurice, 1970; Kottelat, 1997; O'Maoileidigh & Bracken, 1989; Scott & Crossman, 1973; Wright & Giles, 1991.

Author: Andrew M. Hindes, Environment Agency

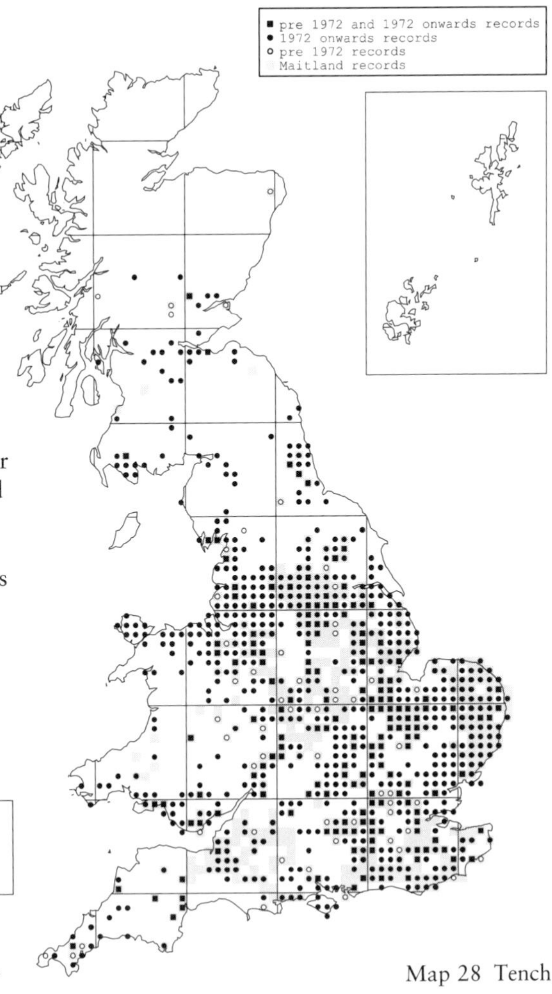

Map 28 Tench

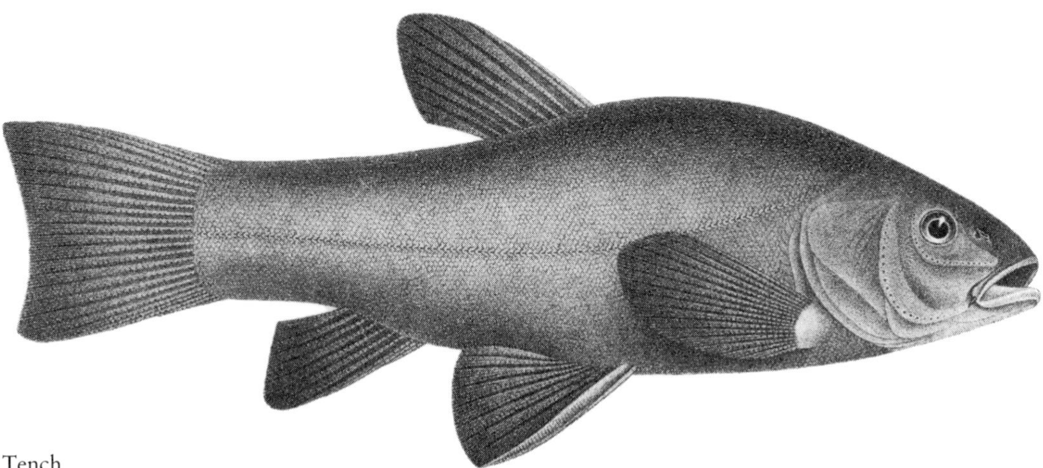

Tench

Spined loach *Cobitis taenia*

Description

- Small, elongate and laterally flattened.
- Small head with low 'forehead' and eyes high on the head.
- Ventral mouth with six small, inconspicuous barbels on upper jaw.
- Small, movable, double spine in small groove below eye (for which spined loach is named).
- Strongly patterned with numerous spots and blocks of dark colour along flanks and with pale underside.

Size

Average length between 5–10cm.

Biology and behaviour

Spawning takes place between April and June with an elaborate tactile courtship involving considerable physical stimulation by the male. Males have a thickened second ray to their pectoral fin forming what is known as the organ of Canestrini, which is thought to be used during courtship and mating. The sticky, yellow eggs are shed among stones or vegetation. After hatching, the juveniles immediately adopt a benthic lifestyle. Females can grow faster than males and reach a larger size. Spined loach appear to be mainly nocturnal and spend most of their time on the bottom or in the substrate and are therefore difficult to find. They have been sampled successfully using electro-fishing and a modified, hand-hauled shrimping trawl.

Spined loach have a relatively high area of gill surface in relation to body weight, an adaptation that aids oxygen uptake, which is probably related to their sedentary, burrowing habits. In addition, spined loach, as with other members of the loach family, are able to take up oxygen directly from air swallowed into the gut.

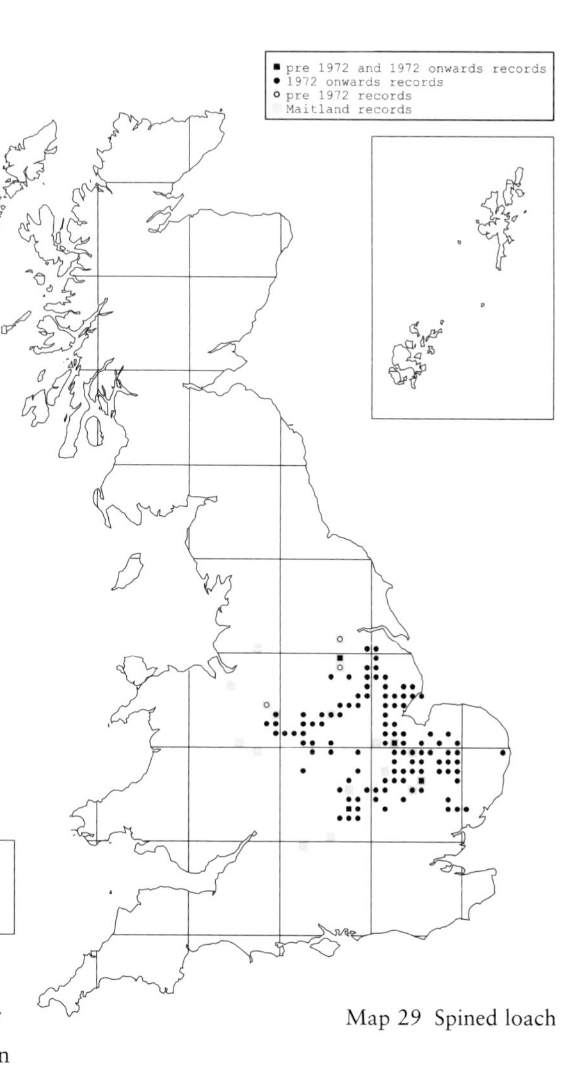

Map 29 Spined loach

Spined loach

Many strong populations of spined loach in Britain have been recorded at sites where levels of dissolved oxygen can fall dramatically in summer, in situations where other fish species would have difficulty surviving.

Habitat

Spined loach have been recorded in a variety of micro-habitats within shallow rivers, streams, drains and ditches and some shallow lakes and gravel pits. A common factor at sites where spined loach occur seems be a fine sand/silt substrate where physical shelter and small food items can be sought. Spined loach have a specialized buccal feeding mechanism by which mud is taken into the mouth, food items (mainly small invertebrates) are passed into the gut and the filtered mud is then spat out.

Distribution in Britain

Spined loach occur naturally in five major river catchments in the east of England: the Rivers Trent, Welland, Witham, Nene and Great Ouse, within which their occurrence is fragmented and localized. The natural distribution of spined loach in Britain is thought to be limited to these river catchments because, following the last glaciation, these rivers drained into a land bridge that connected them with rivers flowing from continental Europe (see Figure 1, p. 19). This connection enabled natural colonization and dispersal of several fish species in these eastward flowing rivers. In the case of spined loach, this distribution pattern remains largely intact today, probably because the lack of any commercial or angling interest in them has served to restrict any deliberate or accidental spread to other catchments. Inevitably there are a few exceptions, such as the introduction of spined loach to the River Stour catchment in Essex/Suffolk and the River Ancholme in Lincolnshire, both brought about through water transfer schemes between river catchments. Comparison of present distribution with that known in 1972 suggests that some of the historical records outside the five known major river systems are probably inaccurate. Spined loach do not occur in Wales or Scotland.

World distribution

The spined loach is extremely widely distributed, occurring from Britain (but not Ireland) in the west across Europe and northern Asia to Japan in the east. In continental Europe the distribution extends from the Netherlands, southern Sweden and the Baltic States in the north, to Spain, Italy and Greece in the south.

Status

- Protected under the Habitats and Species Directive 92/43/EEC and the Bern Convention and listed under the UK Biodiversity Action Plan.

Hybrids and related species

- No hybrids known in Britain.
- The number of *Cobitis* species, subspecies or races in Europe is a matter of debate, with up to 20 distinct species having been described.
- Studies in Japan have described races that occupy different geographic ranges with specific and habitat preferences leading to genetic isolation and speciation.

Spined loach and Man

The spined loach has no commercial or angling value to humans and has therefore been largely ignored. Even its ecology is poorly known compared with most other British freshwater fishes. The listing of spined loach on Annex II of the EC Habitats and Species Directive has led to increased interest in the conservation of the species, including the identification of Special Areas for Conservation (SACs) for it under the Directive. Sites identified as candidate SACs for this interesting and reclusive little fish include Baston Fen, Nene Washes, Ouse Washes and River Mease (Trent Catchment), and it is also an important species at the Fenland SAC, which includes Wicken Fen.

Further Reading: Mann, 1996a; Perrow & Jowitt, 1997, 2000; Robotham, 1978a, b.

Author: Nick Bromidge, Environment Agency

Stone loach
Barbatula barbatula

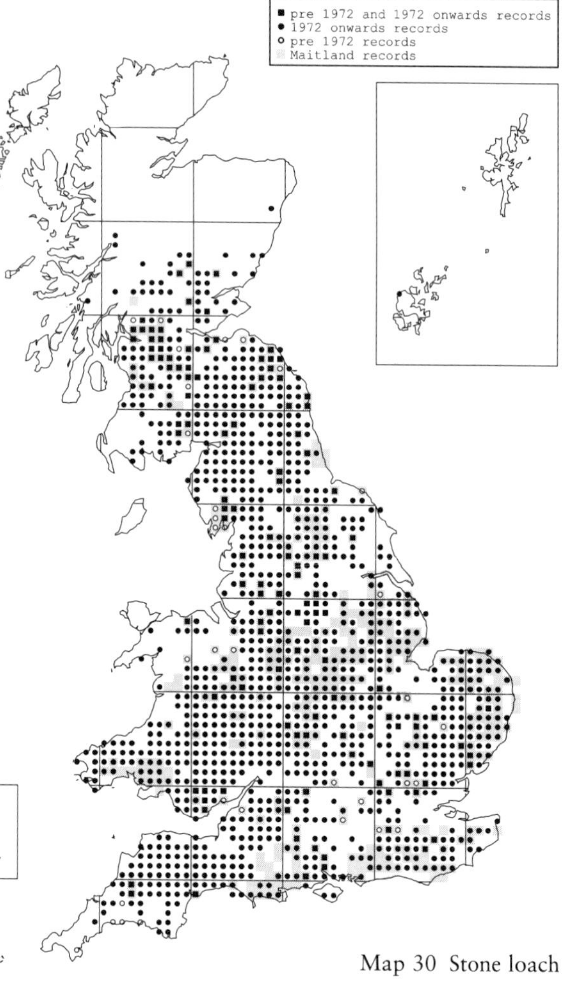

Map 30 Stone loach

Description

- Small, elongate and cylindrical with slightly flattened belly.
- Small head with eyes high on the head; generally longer, flatter, more pointed profile than spined loach.
- Ventral mouth with six prominent barbels on upper jaw.
- Body patterning variable from dark olive to light yellow-brown.

Size

Average length up to 12cm.

Biology and behaviour

Male stone loach are characterized by slightly longer and more pointed pectoral fins than females. During the breeding season males develop minute tubercles on the pectoral fins. Spawning takes place from spring to summer when the sticky eggs are shed among gravel and aquatic vegetation. Stone loach produce large numbers of eggs (*c.* 10,000), with some females shedding the equivalent of almost half their body weight as eggs, laid at intervals throughout the spawning season.

When feeding, a stone loach uses the group of highly developed sensory barbels surrounding the mouth to find and root out small invertebrates that lie buried in sand or between small stones and gravel. Feeding behaviour may involve small groups of stone loach working together over a suitable substrate.

Stone loach

As is common with other loaches, stone loach have the ability to gulp down air into the gut and by so doing absorb oxygen directly through the gut lining. This adaptation may help to sustain them if dissolved oxygen levels fall in the surrounding water.

Habitat

Stone loach are generally found in streams and rivers with a mixture of cobbles, gravel, sand and silt, moderate flow velocities and high organic enrichment, sometimes including the lower reaches of the 'trout' zone. Stone loach spend the daylight hours hiding under small stones and gravel or within stands of dense aquatic vegetation.

Distribution in Britain

Stone loach are found all over mainland Britain except the northern half of Scotland. After the last glaciation, stone loach were probably restricted to catchments in south-east England. They have subsequently spread to most parts of Britain, probably as a result of the former angling practice of using stone loach as live or dead bait. The overall range of the species in England and Wales appears to have changed little in the last 30 years although it is now known from many more 10km squares.

World distribution

Stone loach occur widely across Europe and Asia as far east as China. In Europe they have been introduced to Ireland, but are absent from the Iberian and Italian peninsulas and the northern parts of Scandinavia.

Status

- Not threatened in Britain, in continental Europe or globally.

Hybrids and related species

- No hybrids known in Britain.
- Stone loach now classified in a different family (Balitoridae) to spined loach (Cobitidae). Most Balitoridae species occur in Asia.

Stone loach and Man

The importance of stone loach in the diet of other fishes has long been recognized by anglers and this led to their successful use as bait fish, though they are rarely used as such today. Stone loach were formerly taken for human consumption in Britain, on account of a 'distinctive flavour' and reports of their health-giving properties, and they are still eaten in France. Stone loach nowadays receive attention only from fishery biologists or from children playing in and exploring streams. Stone loach often remain motionless when exposed carefully and therefore fall easy prey to inquisitive youngsters.

Author: Nick Bromidge, Environment Agency

Black bullhead *Ameiurus melas*

also known as American black bullhead catfish

Description

- Superficially similar to the Danube catfish.
- Moderately stout body with broad head, wide mouth and 4 pairs of long but unequal barbels.
- Back dark brown, sides greenish with golden tinge, underside yellow or dull white; dorsal fin membrane black. Males at spawning are jet black above with bright yellow belly.
- Dorsal fin long but short based; adipose fin small, fleshy; anal fin long based; tail fin rounded.

Size

In Europe, adults grow to 20–30cm in length and up to 0.45kg in weight but, in North America, can reach at least double this size.

Biology and behaviour

Black bullhead spawn in the summer, the female producing up to 4000 eggs, with both parents guarding the nest. The fry are guarded by the male until they disperse to feed. The young form schools, spherical in shape, which move through the water like a submerged football. Juveniles feed on small crustaceans, but adults are omnivorous and voracious, favouring benthic organisms, with large specimens predating other fishes.

Habitat

Prefer rich, still waters or sluggish lowland rivers and canals.

Distribution in Britain

Recorded in the wild in Britain from a lake in Warwickshire. The black bullhead is increasingly popular as an ornamental species for garden ponds, which could lead to accidental and intentional, but illegal, introductions to the wild in Britain.

World distribution

Native to North America, black bullhead were introduced into continental Europe in the late nineteenth century, where they have become widespread and are now considered to be a pest in areas where they are naturalized in large numbers, competing with native, bottom-dwelling fishes.

Status

- In common with all other *Ameiurus* spp., listed on the Prohibition of Keeping or Release of Live Fish Orders (see Appendix 3) as a species for which release to the wild is not permitted without a licence.

Hybrid and related species

- Closely related to the American brown bullhead catfish *Ameiurus nebulosus* (Lesueur).

Black bullhead and Man

Originally introduced as an ornamental species, they can be taken by anglers.

Further reading: Wheeler, 1998.

Authors: Sarah Chare & Robin Musk, Environment Agency

Map 31 Black bullhead

Black bullhead

Danube catfish *Silurus glanis*

also known as Wels or European catfish

Description

- Elongated body with smooth scaleless skin coated in mucus.
- Broad, flat head, widely spaced nostrils, small eyes and large mouth with two very long, slender barbels on upper jaw and four short barbels below protruding lower jaw.
- Very short dorsal fin and stubby ventral fins, but long anal fin extends for about half the total length of the fish.

Size

In Britain, grow up to 1.5m long and weigh *c.*20kg, but in continental Europe growing to up to 3m in length and 300kg in weight.

Biology and behaviour

Danube catfish spawn in spring once water temperatures reach 19°C. They reach maturity in five years and may live as long as 30 years. They feed mainly on other fishes, molluscs and crayfish, mainly by scavenging on the bed of a river or lake, although they will rise to the surface to capture fishes, water birds, amphibians and small mammals. Juveniles eat fish fry and invertebrates.

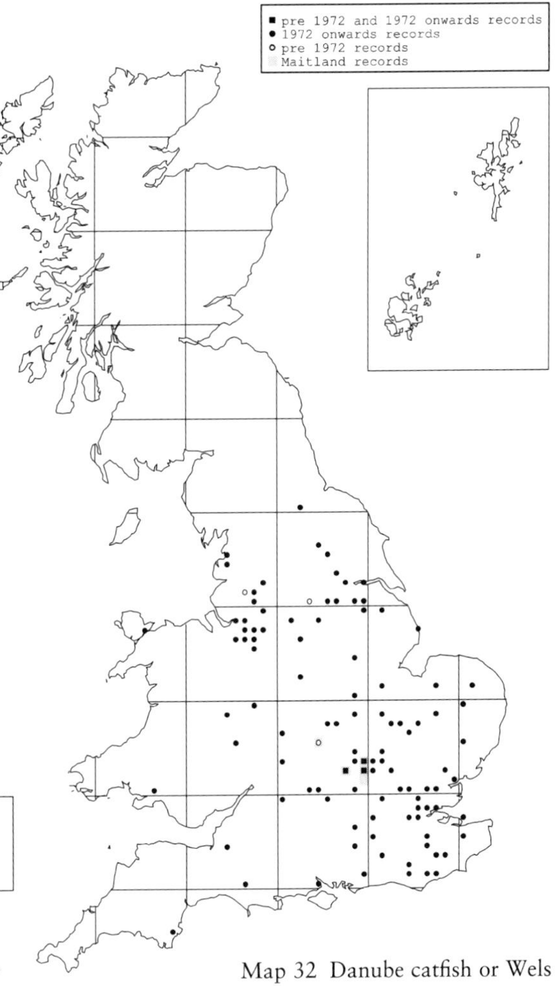

Map 32 Danube catfish or Wels

Habitat

Danube catfish occur naturally in lakes and large flowing rivers, and also sometimes in brackish waters.

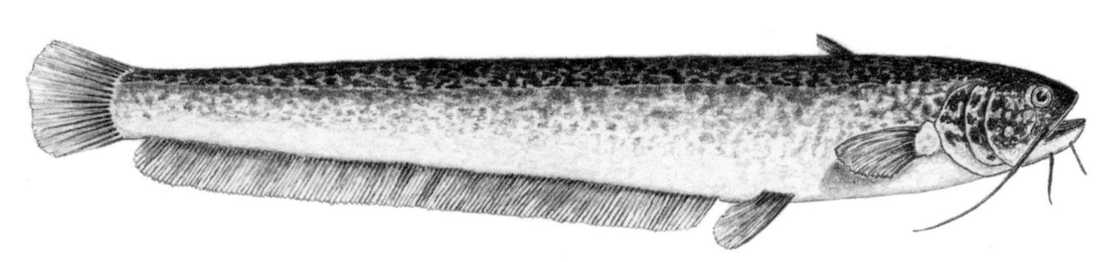

Danube catfish or Wels

Distribution in Britain

Despite some unsubstantiated earlier records, Danube catfish were almost certainly first brought into England from Romania in the 1860s by the Acclimatisation Society. The most successful introduction was that by the ninth Duke of Bedford in 1880 to lakes at Woburn, Bedfordshire, from which further introductions were made. The popularity of Danube catfish with anglers has led them to be introduced to lakes and other still waters throughout England and a few sites in Wales whence their spread into rivers may have resulted from flooding. The increase in records since 1972 is particularly evident.

World distribution

Danube catfish are native to continental Europe and western Asia, from the Danube Basin east to the Aral Sea and the Amudar'ya (known since the time of the Ancient Greeks as the River Oxus) in Central Asia. They have also spread into Germany, France, Spain and northern Italy.

Status

- Listed under the Wildlife and Countryside Act 1981 and, in common with all other *Silurus* species, on the Prohibition of Keeping or Release of Live Fish Orders (see Appendix 3) as a species for which release to the wild is not permitted without a licence.
- Although it is protected under the Bern Convention, this does not apply to Britain because the species is not native here.

Hybrids and related species

- Aristotle's catfish *S. aristotelis* (Agassiz), found in Greece, is the only other species of the catfish family (Siluridae) native to Europe.

Danube catfish and Man

The steady increase in demand among coarse anglers for unusual, large specimen fishes has meant that Danube catfish have been stocked illegally into waters in England and Wales. In addition to being a voracious predator, Danube catfish can carry the notifiable disease of carp, spring viremia. In Hungary, they are farmed for the table and also appear on menus in Austria.

Further reading: Lever, 1977.

Authors: Sarah Chare & Robin Musk, Environment Agency

Pike *Esox lucius*

Description

- Unmistakable, streamlined appearance with all power at the rear.
- Elongate head and very large mouth, with lower jaw angled backwards and well-developed, needle-sharp teeth.
- Single dorsal fin almost aligned with the anal fin.

Size

Adults normally range from 40–100cm in length and 2–14kg in weight. However, specimen fishes up to 19kg and measuring 130cm have been caught in Britain.

Pike

Biology and behaviour

Males and females normally mature in two or three years but, in ideal conditions for growth, spawning can occur after one year. Spawning takes place in shallow water and the eggs are shed indiscriminately over submerged and emergent water plants. Pike are solitary but not territorial, though young individuals may have a limited home range. They prey on almost any living thing depending on its size, abundance and seasonal availability: invertebrates (mainly eaten by young pike), fishes and other vertebrates. Pike are normally considered to be fast growing, but the availability and size of prey fishes in different habitats and the population density of pike may influence growth rates.

Habitat

Pike are capable of living and thriving in many different waters, from lakes to larger ornamental ponds and from canals and slow-flowing rivers to streams. They are able to survive in waters with some pollution and low oxygen content. Although they can tolerate very cold temperatures, high temperatures (above 29°C) can prove fatal. Some diversity of habitat structure may be important, particularly for young pike which are often associated with weed beds in shallow, sheltered areas. Larger pike are usually associated with deeper, more turbid waters.

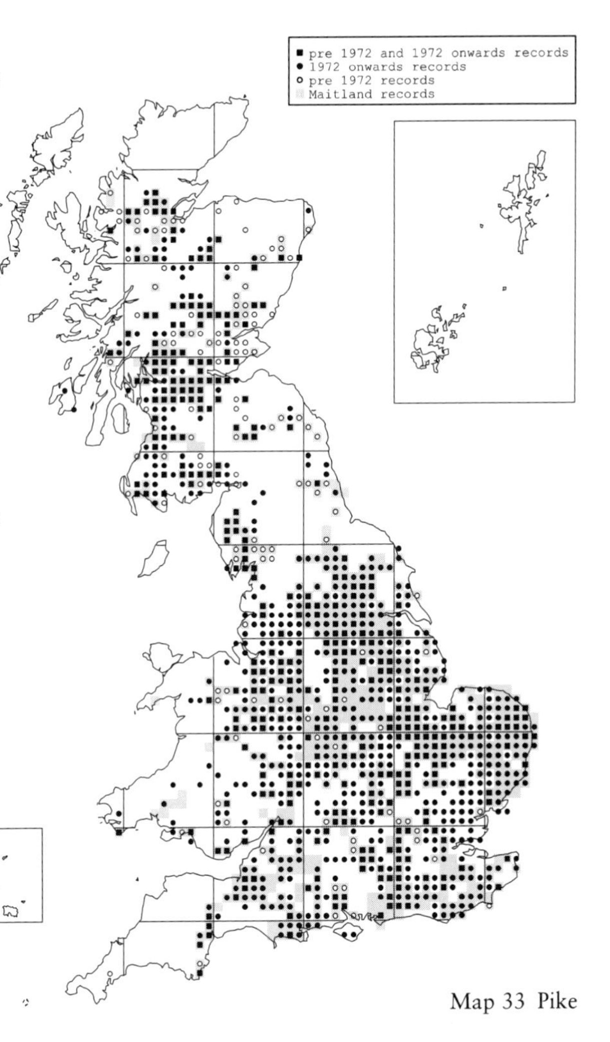

Map 33 Pike

Distribution in Britain

Pike are found throughout most of Britain, but are absent from large areas of northern Scotland and most of the offshore islands (except Islay), and many upland areas elsewhere. Pike may not be native throughout this range, possibly having moved or been moved north and west, away from a 'native' area in south-east England.

The absence of pike in many upland areas is probably due to failure to colonize fast-flowing rivers and isolated water bodies. Pike have been introduced to some sites in the uplands where they survive as self-maintaining populations.

There has probably been little change in the distribution of pike since 1972. The apparent increase in range in lowland areas since that date (such as in eastern Essex and Suffolk, Hertfordshire, Gloucestershire and Wiltshire) is misleading because these areas were poorly covered for many common species in the 1972 maps.

World distribution

Pike are probably the most widely distributed piscivorous fishes in the northern hemisphere. They occur in Asia and North America and throughout much of Europe, but are absent from parts of the Iberian Peninsula, southern Italy, the southern Balkans and the Mediterranean islands. It was probably introduced to Ireland in the fifteenth century.

Status

- Not threatened in Britain, in continental Europe or globally.

Hybrids and related species

- No hybrids known in Britain.
- Only European species of pike.

Pike and Man

Pike has many local names and is a popular sport fish for anglers. It has palatable flesh, but many small bones. Pike was a popular but scarce and expensive food item in mediaeval England, but is now rarely eaten in Britain. Its importance in mediaeval and Tudor times may be reflected in the fact that the pike (termed *lucy* or *luce*) appears as a heraldic device in some coats of arms. In Chaucer's *Canterbury Tales*, written at the end of the fourteenth century, it was a measure of the wealth of the Franklin, a rich landowner, that he had 'many a luce in stew', meaning many pike in his fish pond. Pike are still farmed commercially for food in parts of continental Europe, and the flesh is the essential ingredient of the classic French dish *quenelles de brochet*.

Authors: Paul Harding, Centre for Ecology and Hydrology, & Matt Carter, Environment Agency.

Smelt *Osmerus eperlanus*

also known as Sparling, particularly in Scotland

Description

- Slender fish with almost translucent skin and large scales and eyes.
- Top of head and back dark grey-green; sides silvery green.
- Relatively large mouth with projecting lower jaw and numerous long, sharp teeth on both jaws.
- Fleshy adipose fin between the dorsal fin and caudal fin.
- Unusual, characteristic smell of cucumber.

Size

Length normally 10–20cm, sometimes up to 30cm and over 150g in weight. Males usually smaller than females.

Biology and behaviour

During October and November smelt often form pre-spawning shoals in estuaries and around river mouths; between February and April they ascend the rivers to spawn. Water temperature is thought to be the main cue for timing of the spawning run, with temperatures ranging from 4°–7°C initiating the run. However, the tidal cycle and river flows are also thought to be important because smelt are not particularly strong swimmers.

Spawning normally occurs in sand and gravel substrates in fast-flowing water, around the upper limit of tidal influence. Because the

Smelt or Sparling

spawning run of smelt takes place over a relatively short period, usually only a week, spawning sites are often difficult to detect. Each female lays between 10,000 and 40,000 eggs, which are initially pale yellow and about 1mm in diameter. Normally, many adults die after spawning, but some survive to return to the sea and spawn in subsequent years. Smelt that spawn more than once can live for up to six years.

Hatching normally occurs in 20–35 days, but this depends on the water temperature. On hatching, the smelt fry move down into the estuaries where they feed on small zooplankton such as protozoans and rotifers. As they get bigger they prey on increasingly larger items, such as planktonic crustaceans (including shrimps and mysids) and young fishes.

Habitat

Smelt normally inhabit unpolluted estuaries, except during spawning when they move into fresh water, but only as far as the upper limit of tidal influence.

Distribution in Britain

Smelt were formerly present all around the coast of Britain in large estuaries, but they have disappeared from many probably because they are susceptible to pollution and other human activities (including fishing). For example, they were previously recorded from at least 15 rivers in Scotland, but have now disappeared from all but the Rivers Cree, Forth and Tay. In recent years, improvements in water quality have seen smelt return to a few English rivers from which they had gone, such as the Thames and Trent. Populations are also known in the Rivers Humber, Nene, Great Ouse and the Waveney at Breydon Water in Norfolk. A non-migratory freshwater population was once present in Rostherne Mere in Cheshire, but this became extinct in the 1920s, probably as a result of deteriorating water quality.

World distribution

Smelt occur along the coasts of Europe from north-western Spain in the south, to Scandinavia and western Russia in the north. They are tolerant of a wide range of salinities and in Scandinavia there are several non-migratory freshwater populations.

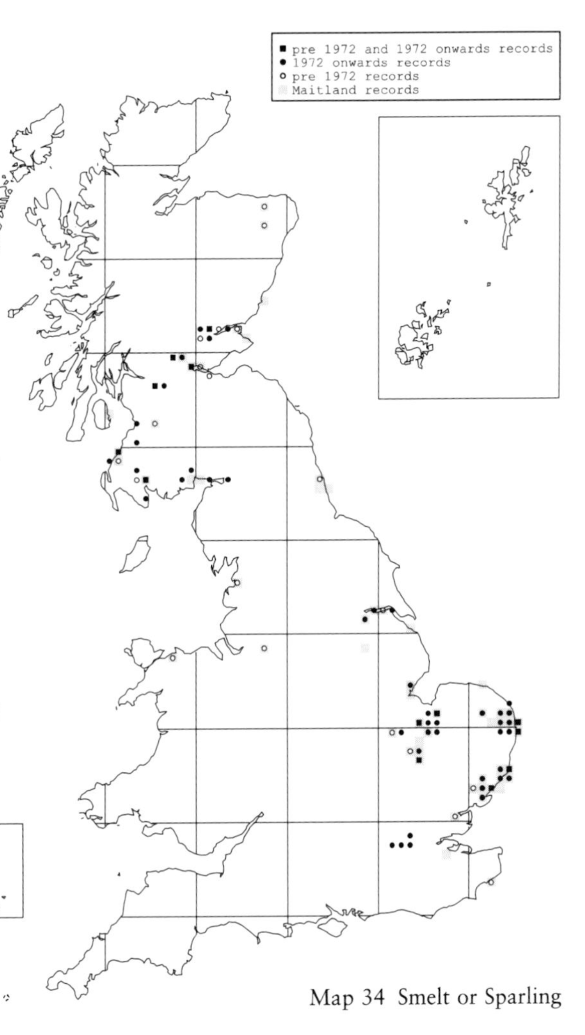

Map 34 Smelt or Sparling

Status

- Despite their apparent vulnerability, smelt are not protected under British or European legislation.

Hybrids and related species

- No hybrids known in Britain.
- No other species of smelt occur in Europe, although there are ten related species world-wide. Capelin *Mallotus villosus* (Müller), which is found in huge numbers in the Arctic seas, is closely related.

Smelt and Man

In the late nineteenth century smelt supported valuable commercial fisheries in Britain and provided an important source of revenue to local netsmen. Commercial smelt fisheries rely on the vulnerability of the species to capture in traps and nets during their short spawning run. This vulnerability, together with restrictions on access to and the destruction of spawning grounds, has contributed to their shrinking distribution. They are sold for consumption, either fresh or smoked, and are also used by pike anglers as a very effective dead bait.

Further reading: Ellison & Chubb, 1968; Hutchinson & Mills, 1987; Maitland, 2003; Maitland & Lyle, 1991, 1996; Winfield *et al.*, 1994.

Author: Mike Atkinson, Environment Agency

Vendace *Coregonus albula*

Description

- Streamlined and slender, with overall white appearance encompassing a deep blue-green or blue-grey back, graduating through silver to a white underside.
- Small adipose fin.
- Smaller than other coregonids with which it might be confused, but leading lower jaw is a distinguishing feature.

Size

Adults in Britain normally measure up to 25cm in length and weigh 150g.

Biology and behaviour

Spawning takes place from late November to mid December, when males gather after dusk in the spawning areas as described below. Females then join them, each bearing thousands of eggs 1–2mm in diameter. A redd is not constructed and the eggs are simply scattered on the lake bed, where they are fertilized. The young hatch out mainly during the following March and April, the timing depending on water temperatures over the course of the winter.

Vendace shoal throughout their life-cycle, although at night these aggregations become looser when individuals begin to feed. The diet of all age classes is dominated by zooplankton, particularly cladocerans such as water fleas (*Daphnia* spp.). At times, adult vendace may also take insect larvae and pupae, in particular those of non-biting midges (Diptera: Chironomidae).

Habitat

Although some vendace populations in northern Europe migrate to the brackish waters of the Baltic Sea, both British populations are resident in lakes throughout their life-cycles. They require relatively cool water with high oxygen levels and typically spend the daylight hours offshore at depths in excess of 10m, where they lie very close to the lake bed. At night, however, they move up the water column and may even break the

Vendace

offshore water surface at times. Vendace also require a spawning habitat with a clean substrate, which is typically an inshore area of gravel at depths of less than 4m.

Distribution in Britain

Now restricted to Bassenthwaite Lake and Derwent Water in the English Lake District, the two former populations in south-west Scotland, in Castle Loch and Mill Loch, having become extinct during the last century. The genetics and morphology of the two surviving populations are very similar and it is possible that individuals occasionally move between the two lakes along the connecting River Derwent.

In recent years, attempts have been made to reintroduce vendace to Scotland through the establishment of new populations in Daer Reservoir and Loch Skeen, by the introduction of young hatched from eggs collected from Derwent Water and Bassenthwaite Lake respectively. Whether the introductions have been successful and self-sustaining populations established at either site, has not yet been determined.

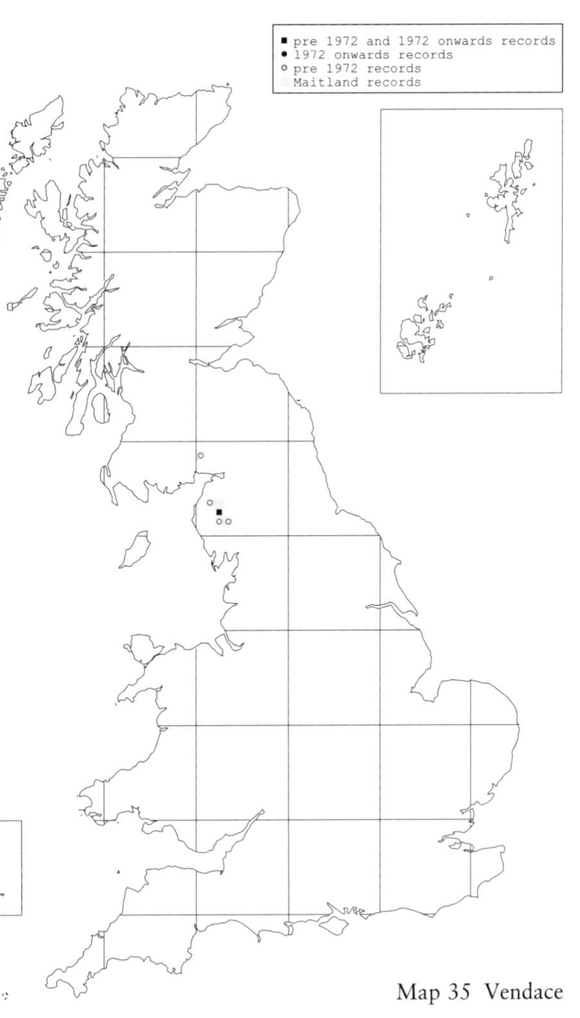

Map 35 Vendace

World distribution

Widespread in northern Europe, particularly in Finland and other parts of Scandinavia. Vendace have also been introduced to some lakes in central Europe for the development of commercial fisheries.

Status

- Protected under the Habitats and Species Directive, Bern Convention and Wildlife and Countryside Act and listed under the UK Biodiversity Action Plan.
- Threatened in parts of its European range by factors including eutrophication, the silting up of spawning grounds, lake level fluctuations, species introductions and over-exploitation.

Hybrids and related species

- No hybrids known in Britain.
- Related most closely to other coregonids: whitefish and houting, and to pollan *Coregonus autumnalis* which occurs in Ireland but not in Britain.
- More distantly related to game species including Atlantic salmon, brown and sea trout, Arctic charr and grayling.

Vendace and Man

The vendace has a delicate flavour and its flesh is rich in oils. It is heavily exploited in commercial fisheries in many parts of Europe, particularly in Finland, where it is taken principally in gill nets but also in trawl and seine nets. In Finland it is frequently eaten raw. In Britain, however, commercial fishing has always been extremely limited; only catches from the two now-extinct populations of south-west Scotland were ever documented. Commercial and recreational fishing for this species is now banned completely in Britain by conservation legislation.

Humans continue to threaten British vendace populations by the illegal introduction of new fish species to their lakes in the form of escaped or discarded live-bait used for pike fishing. The species causing greatest concern is the ruffe, which has been introduced to both Bassenthwaite Lake, where it feeds extensively on vendace eggs, and more recently to Derwent Water.

Further reading: Winfield *et al.*, 1994.

Author: Ian J.Winfield, Centre for Ecology and Hydrology, Lancaster

Whitefish *Coregonus lavaretus*

also known as Schelly in England, Gwyniad in Wales, and Powan in Scotland

Description

- Streamlined, but rather heavily built, with overall white appearance encompassing a blue-grey back graduating through grey to a whitish underside.
- Small adipose fin.
- Can be confused only with other coregonids, in particular houting, but the leading upper jaw of whitefish distinguishes it from the vendace and is much less pronounced than the 'snout' of houting.

Size

Maximum length typically 30–35cm and weighing up to 400g.

Biology and behaviour

Spawning takes place from late December to early February, with some variation among the separate British populations. Like the vendace, males gather after dusk in the spawning areas and are joined by females, each bearing thousands of eggs 2–3mm in diameter, which they scatter on the lake bed where they are fertilized. Incubation takes about 90–100 days and they hatch mainly during the following March and April, depending on winter temperatures.

Whitefish shoal throughout their life-cycle, although at night these aggregations become looser when individuals begin to feed. The diet of young whitefish is usually dominated by zooplankton, particularly cladocerans such as water fleas (*Daphnia* spp.), but adults may feed almost exclusively on macroinvertebrates of rocky inshore areas including water hoglice (*Asellus* spp.) and freshwater shrimps (*Gammarus* spp.).

Habitat

All British populations are resident in lakes throughout their life cycles, unlike some whitefish populations in northern Europe which migrate to sea[1]. In Britain they require cool water with relatively high oxygen levels. They typically spend daylight hours offshore at depths in excess of 20m. At night they usually move into inshore areas to feed in water as shallow as one to four metres deep . Whitefish also require a spawning habitat containing a clean substrate, which is typically found in inshore areas of gravel at depths of less than four metres.

[1] See notes on houting (p. 37)

Whitefish or Powan, Schelly or Gwyniad

Distribution in Britain

Whitefish are restricted to Brotherswater, Haweswater, Red Tarn and Ullswater in the English Lake District, Loch Eck and Loch Lomond in Scotland, and Llyn Tegid (Bala Lake) in Wales. The genetics and morphologies of these seven populations show considerable variation, notably the relatively larger eye of the gwyniad of Llyn Tegid.

Recently, attempts have been made to establish refuge populations in England (Blea Water and Small Water from the Haweswater population) and Scotland (Carron Reservoir and Sloy Pond from the Loch Lomond population) while similar action is currently being considered in Wales. Whether a self-sustaining population has been established from any of these introductions is not yet known.

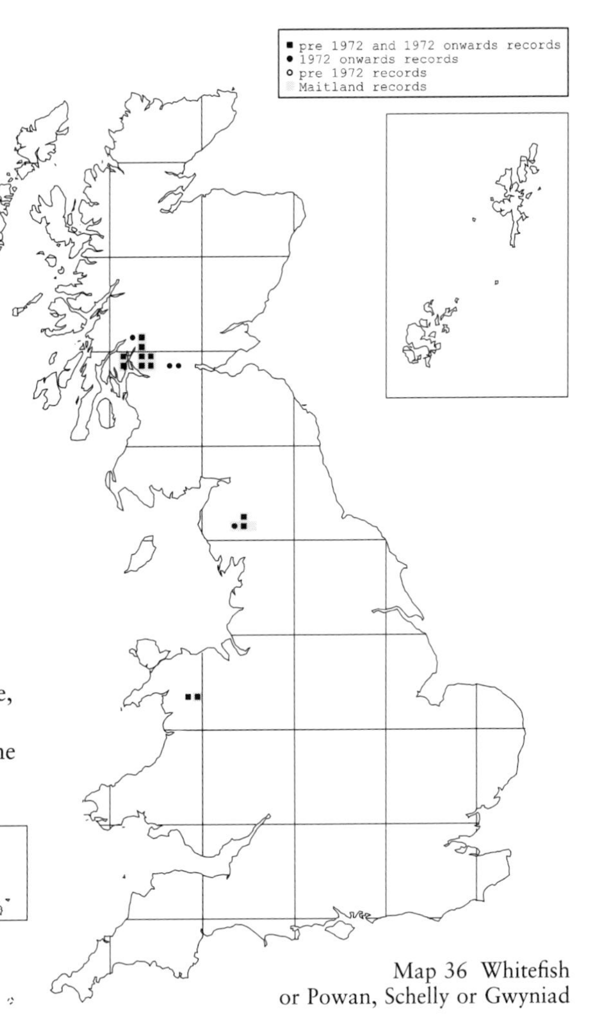

Map 36 Whitefish or Powan, Schelly or Gwyniad

World distribution

The whitefish is widespread in northern Europe, particularly in Norway and other parts of Scandinavia. In addition, it also dominates some Alpine lakes in central Europe, although many such populations are the result of past stockings aimed at developing commercial fisheries. As the result of extensive historic movements and translocations of fish, together with subsequent hybridizations, many current populations cannot be identified beyond the generic level.

Status

- Protected under the Habitats and Species Directive, Bern Convention and Wildlife and Countryside Act, and listed under the UK Biodiversity Action Plan.
- Threatened in parts of its European range by factors including eutrophication, siltation of spawning grounds, lake level fluctuations, species introductions and over-exploitation.

Hybrids and related species

- No hybrids known in Britain.
- Closely related to other British coregonids, vendace and houting, and to pollan *Coregonus autumnalis*, which occurs in Ireland but not in Britain.
- More distantly related to game species including Atlantic salmon, brown and sea trout, Arctic charr and grayling.

Whitefish and Man

Like vendace, whitefish are reported to have a delicate flavour; their eggs can be used as a form of caviar. The species is heavily exploited in commercial and recreational fisheries in many parts of its European range, where it is caught in gill, trawl, and seine nets, in traps and by anglers. In many lakes in continental Europe the population ecology of whitefish is artificial, being determined largely by the annual rates of stocking and exploitation. In Britain, in contrast, fisheries have always been extremely limited and are now banned completely by conservation legislation.

As with vendace, British whitefish populations continue to be threatened by the introduction of new fish species to their lakes. Such introductions are illegal in England and Wales (but not yet in Scotland). The ruffe is of particular concern as it has now been introduced to both Llyn Tegid and Loch Lomond where it feeds extensively on whitefish eggs.

Further reading: Winfield *et al.*,1994.

Author: Ian J. Winfield, Centre for Ecology and Hydrology, Lancaster

Rainbow trout
Oncorhynchus mykiss

Description

- Streamlined, although fish farm strains are usually deeper-bodied.
- Characterized by multiple colours and prominently spotted tail, commonly with wide band of red, pink or mauve along flank.
- Stream residents and spawning males tend to be darker; lake residents and sea-migrants more silvery until attaining maturity.

Size

Selective breeding means that there is no average size. Some British rod-caught specimens have exceeded 15kg in weight.

Biology and behaviour

The farmed strains that are prevalent in Britain tend to grow fast, mature in their second or third winters and die young at three to four years. In contrast, wild strains grow less rapidly initially, mature at older ages and live longer. In Britain, spawning takes place from November but more commonly from February to May in gravel beds in streams or still waters, where the eggs are laid in redds and then covered. Few of the juveniles appear to survive in Britain, although eggs can be fertilized artificially and incubated successfully over a wide range of water temperatures (0°–13°C). Wherever both rainbow and brown/sea trout species have been naturalized worldwide, with increasing environmental temperature rainbow trout tend to predominate. However, in the more equable British climate, rainbow trout usually fail to thrive in competition with native salmonids. This is possibly because the eggs of rainbow trout hatch later and insufficient fry are able to gain feeding territories. Although they grow well in the wild, long-term husbandry of hatchery strains may result in loss of competitive vigour and fitness for life in the wild, and further limit success in natural reproduction.

Rainbow trout are usually considered to range more actively when feeding than brown/sea trout with which they occupy many habitats in common. They feed on insects, small crustaceans and other fishes, depending on availability. The reported maximum size of rainbow trout in the world is 26kg. In Britain,

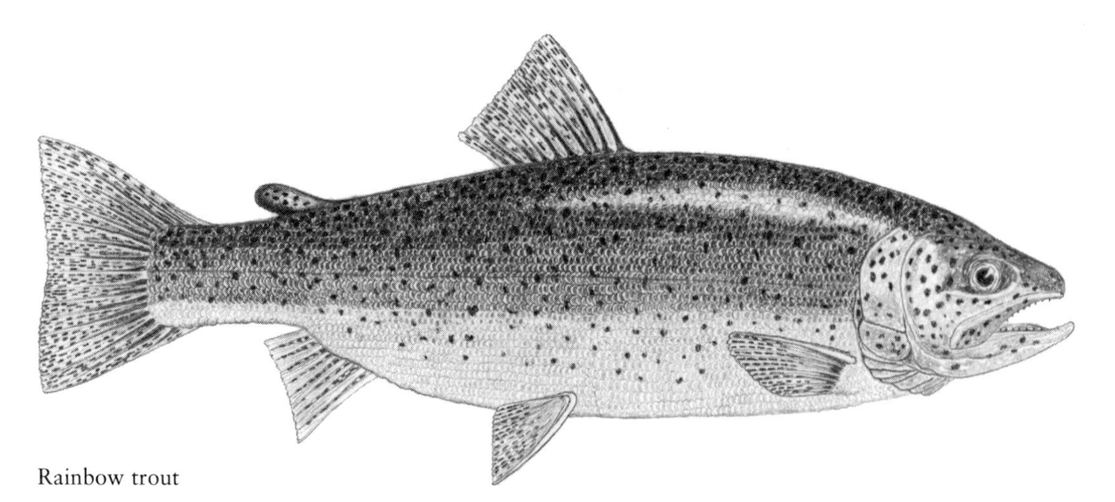

Rainbow trout

rainbow trout are normally stocked at lengths greater than 40cm for put-and-take angling, but they can be reared to weights in excess of 10kg.

Distribution in Britain

Introduced to Britain as a sport fish in the 1880s, rainbow trout were found to be tolerant of intensive rearing and began to be farmed extensively and stocked in still water put-and-take fisheries from the 1970s. The results of a recent questionnaire survey indicate that there are more than 300 stocked fisheries in Scotland alone. These fisheries are common in low-lying areas, near to centres of population, but are still relatively uncommon in the Grampian and the western Highlands and Islands. Rainbow trout that have escaped from fish farms or migrated from stocked fisheries were reported from 48 Scottish rivers. The survey also recorded spawning occurring in 51 localities, in both running and still water, but with very limited success. However, with increasing use of all-female and triploid stocks in recent years, spawning may be less common now than it was when mixed-sex populations were the norm in farms and fisheries.
Throughout Britain, self-maintaining populations were never common, but appear now to be restricted to a very few localities in England, including the Derbyshire Wye.

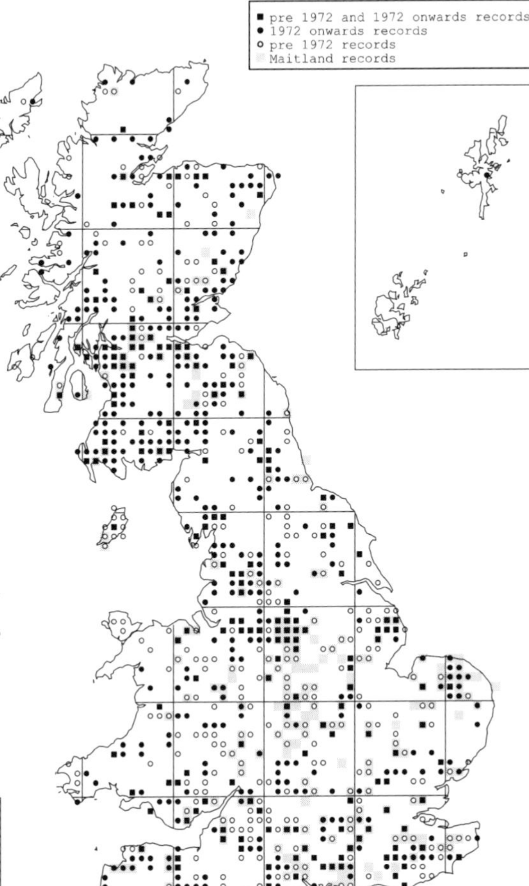

Map 37 Rainbow trout

World distribution

Rainbow trout are indigenous to temperate western North America, where they inhabit cool streams and lakes, or migrate to the ocean in the steelhead form. They have been introduced widely around the world and are present in every continent except Antarctica. In the tropics they are restricted to areas above 1200m.

Status

- Introduced to Britain, but there are few self-maintaining populations.

Hybrids and varieties

- Hybridization of rainbow trout with Atlantic salmon or brown/sea trout is very rare, although it can be achieved artificially, e.g. the brownbow trout.
- Many colour forms of rainbow trout exist, mostly derived from selective breeding.

Rainbow trout and Man

As a result of widespread introductions, rainbow trout are now the most cosmopolitan of the salmonid species. They are extensively farmed for the table, both in fresh water and in the sea, and are highly valued for sport in recreational fisheries, in temperate regions and spring-fed waters throughout the world. Almost 16,000 tonnes were farmed in Britain during 2000, comprising 11,193 tonnes for the table and 4,500 tonnes for restocking for angling.

Further reading: Froese & Pauly, 1998; Frost, 1974; Lever, 1977; MacCrimmon, 1971.

Author: Andy Walker, FRS Freshwater Laboratory

Atlantic salmon *Salmo salar*

Description

- Streamlined, with small, dark adipose fin and narrow 'wrist' at base of tail.
- Back blue-grey, flanks silver with black spots (mostly above the lateral line) and belly white.
- Spawning females (hens) darker in colour, spawning males (cocks) reddish brown with pronounced hook (kype) on lower jaw.
- Parr (juveniles) brownish with dark markings along the flanks and spotted black, brown and red; smolts are silvery in appearance.

Size

Adults returning to fresh water vary greatly in size, normally in the range 40–75cm in length weighing 1–5kg, but occasionally considerably larger.

Biology and behaviour

Spawning takes place in clean, flowing water with a gravel or cobble substrate. The female excavates a depression (redd) by turning on her side and making a series of vigorous undulating motions with her tail. Sea-run males fight to join the female, the dominant male defending the redd from competitors. The female lays her eggs in the redd, where they are fertilized by the sea-run male. Sometimes a mature male parr, a 'sneaky young male' will dart in and fertilize the eggs before the male that constructed the redd gets a chance. The eggs in the redd are covered with gravel by further sweeps of the female's tail. This process may be repeated several times. Spawning occurs between October and January, but usually in November and December.

Juvenile salmon are territorial, feeding on aquatic and terrestrial invertebrates. They remain in fresh water for between one and eight years (commonly one to three years in Britain) before migrating to sea as smolts. They remain at sea, generally for between one to three years, feeding on larger zooplankton and fishes including capelin *Mallotus villosus* (Müller) and sandeels (Ammodytidae), before returning to fresh water to spawn. Salmon have a strong homing instinct and the great majority of fishes return to spawn in the river in which they hatched. Atlantic salmon generally spawn only once, because the majority of fish die after spawning.

Habitat

Juvenile salmon require shallow, clean, well-oxygenated, fast-flowing waters with a cobble and gravel substrate. Habitat with abundant cover, provided by the substrate and to a lesser extent by undercut banks and aquatic vegetation, is favoured. Salmon parr are also well adapted to swimming in riffles. Older parr prefer deeper water, to which fry may also migrate during winter. Adult salmon require access to their spawning grounds and to holding areas that provide refuge on their upstream migration. They are unable to tolerate high water temperatures (greater than 22°–23°C) without acclimation.

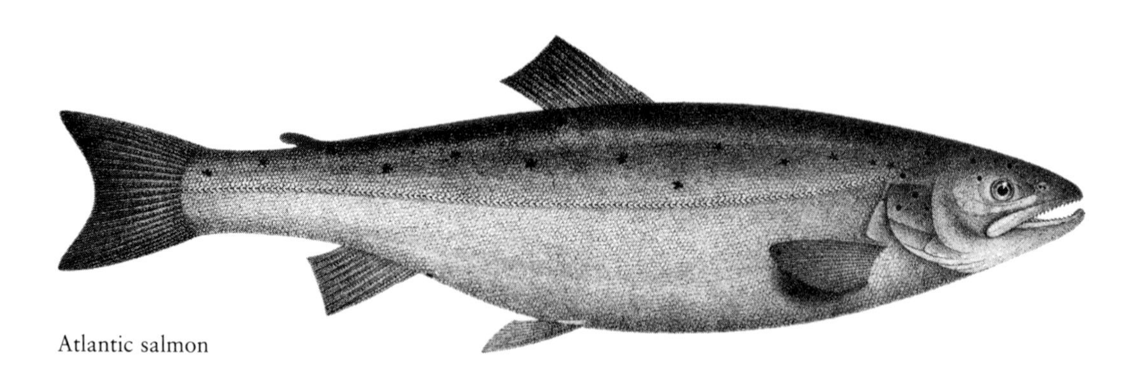

Atlantic salmon

Distribution in Britain

The Atlantic salmon is the only native 'salmon' in Britain, all other 'salmon' being introduced species and, in most cases, not belonging to the genus *Salmo*. It is widely distributed in northern and south-western England, Wales and Scotland. Salmon are also present in the rivers and chalk streams of the south coast of England. They are absent from East Anglia and large areas of central and southern England because of unsuitable natural habitat, habitat degradation, poor water quality and obstructions to the natural passage of fishes in rivers.

Pollution removed salmon stocks from many urban and industrialized rivers, but in recent years improvements in water quality have seen the recovery of salmon populations in many rivers. Comparison of recent records with the distribution in 1972 shows a broadly similar pattern, although there are now many more records, particularly from their stronghold areas.

Map 38 Atlantic salmon

World distribution

Atlantic salmon occur in rivers on the eastern coast of North America, Greenland, Iceland, Scandinavia, the British Isles and western continental Europe as far south as Portugal. They spend the majority of their adult life at sea, feeding in the North Atlantic around Greenland and the Faroe Islands. Several landlocked populations of Atlantic salmon exist in Russia, North America and Scandinavia. Landlocked Atlantic salmon have also been introduced to New Zealand.

Status

- Protected, particularly against over-exploitation, under the Habitats and Species Directive and Bern Convention, and listed under the UK Biodiversity Action Plan.
- Threatened throughout its range by pollution, over-exploitation, habitat degradation, lack of access to spawning grounds and reduced survival rates when at sea.
- Under the Prohibition of Keeping or Release of Live Fish Orders (see Appendix 3) non-anadromous varieties of *Salmo salar* may not be released to the wild without a licence.

Hybrids and related species

- Occasionally hybridizes with trout *Salmo trutta*.
- Related most closely to other species of *Salmo* (trout), also to many species of Salmonidae including salmon and trout other than *Salmo* species, huchen *Hucho hucho* (Linnaeus), charr and grayling.

Salmon and Man

Salmon feature strongly in our culture and history (see Chapter 1), and have long been regarded as making excellent eating. Some commercial fisheries still exploit wild stocks, although marine salmon farms now provide many more fishes for the table. Salmon angling remains a popular and economically important activity, reflecting the fine sporting qualities and high value of the 'King of fishes'.

Salmon forms the basis for many recipes, including preserved as *gravadlax* and smoked salmon, or cooked as *saumon en croûte*; but there is no better dish than a wild salmon, lightly poached and served cold with cucumber and fresh mayonnaise.

Further reading: Crisp, 2000; Mills, 1991; Shearer, 1992; Youngson & Hay, 1996.

Author: Jonathan Shelley, Environment Agency

Brown trout and Sea trout *Salmo trutta*

Description

'Brown trout' is the common name for the species *Salmo trutta*. In this account, the term 'brown trout' is used to describe individuals of *Salmo trutta* that spend their entire life in rivers or still waters, as distinct from the 'sea trout' (a type of brown trout) that start life in freshwater, but which later migrate to sea.

An earlier classification identified not one but ten species of brown trout in the British Isles – including several sea-going forms! This reflected broad physical and behavioural differences in populations around the country. Later work has shown that marked differences can occur in the genetic make-up of *Salmo trutta*, particularly where populations have been reproductively isolated, for example by a physical barrier such as a waterfall. However, these differences are not sufficient to consider the brown trout as several species. Sea trout, for example, are known to breed with resident brown trout when the two come together on the spawning grounds.

The following general features apply to both groups of *Salmo trutta*:

- Streamlined, with blue-grey, olive-brown or black back, silver or golden flanks, with numerous red or brown spots mostly above the lateral line.
- Belly white, silver or golden-yellow.
- Small, dark, red-orange adipose fin, thick 'wrist' at base of tail.
- Spawning females become darker; spawning males reddish brown with pronounced hook (kype) on lower jaw.

The first sign of a difference between brown trout and sea trout occurs as sea trout prepare for the seaward journey, changing cryptic coloration for the bright silvery appearance of the smolt. On return to fresh water, adult sea trout have the following general features:

- Streamlined, with blue-grey back, silver flanks with numerous black spots, mostly above the lateral line, and white belly.
- Small, dark adipose fin, thick 'wrist' at base of square-ended tail.

Size

The maximum size that brown trout or sea trout can attain varies considerably between populations, influenced by genetic and environmental factors. A brown trout resident in an upland stream may not reach even 500g in weight, but the current rod-caught record for a wild brown

Brown and Sea trout

trout, captured from a deep Scottish loch, is over 14kg. Sea trout tend to be larger on average because they can exploit rich feeding available in the sea. The British rod-caught record for a sea trout is currently close to 13kg.

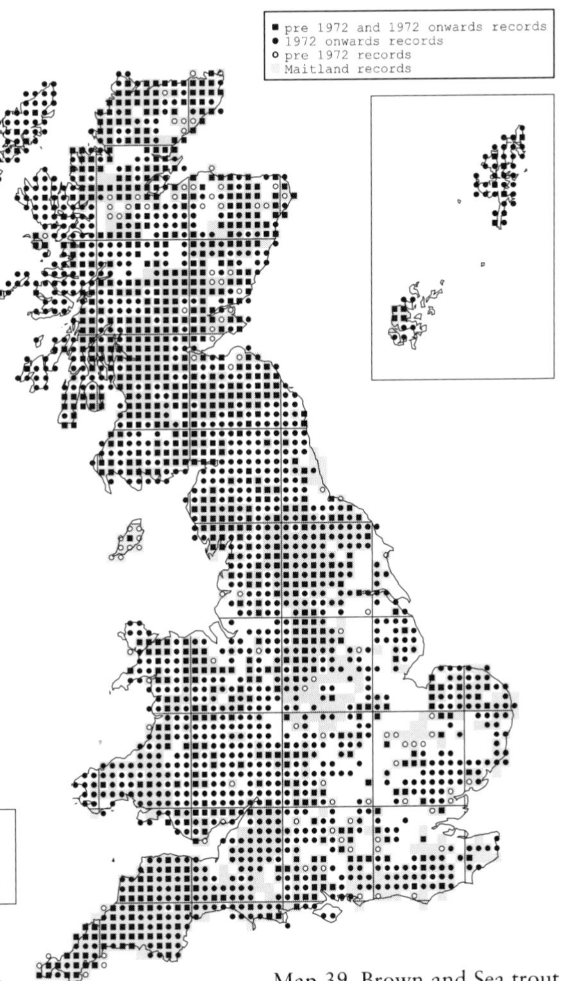

Map 39 Brown and Sea trout

Biology and behaviour

There is no difference between brown trout and sea trout in the process of spawning, and, where both groups overlap, inter-breeding can take place. Spawning usually occurs in clean flowing water with a gravel or cobble substrate, normally between October and December. However, some lake-dwelling populations of brown trout are known to utilize wave-washed, gravel shorelines for spawning if access to in-flowing streams is restricted.

The act of spawning is the same for both sea and brown trout. A female excavates a depression (redd) by turning on her side and making a series of vigorous undulating motions with her tail. The size and depth of the redd and the preferred substrate depends on the size of the female. Males may fight to join a particular female, the dominant male defending the redd from competitors. The female lays her eggs in the redd, where they are fertilized by the male and then covered with gravel by further sweeps of the female's tail. This process may be repeated several times. The total number of eggs laid will also depend on the female's size – averaging around 1800 eggs per kilogram of body weight.

Brown trout may spawn repeatedly over several years and have been reported to live for up to 24 years in Britain – and even longer elsewhere. Sexual maturity is usually attained at the age of two or three years. Populations can undergo various degrees of migration during their life time, some remaining in their natal stream, others moving downstream to reside elsewhere in the river, in still waters or even estuaries (so called 'slob' trout). Trout generally return to their natal river to spawn. The eventual size that fish attain depends on genetic and environmental factors. At the upper extreme, 'ferox' brown trout, which inhabit deep lakes and prey entirely on other fishes, can weigh up to 14kg. Brown trout and pre-smolt sea trout are territorial, feeding on aquatic and terrestrial invertebrates.

The age at which sea trout emigrate to sea as smolts appears to increase with latitude because northern populations experience cooler temperatures and, as a consequence, grow more slowly. In Britain, most sea trout emigrate as smolts after two or three years in fresh water, some after only one year. The smolt migration usually takes place in spring. Unlike salmon, most sea trout are thought to inhabit areas of the sea relatively close to their home rivers (within 150km), but some stocks in the North Sea may range more widely (up to 750km from the natal river). At sea they have a varied diet including crustaceans and other fishes such as herring

Clupea harengus Linnaeus and sandeels (several species of Ammodytidae).

A variable, but often substantial proportion of adult sea trout returning to the river are fishes that have spent only a few months at sea. These are known by a variety of local names, most commonly whitling, herling or finnock. They have an average weight around 500g, with peak runs normally occurring in July or August. Not all whitling appear to mature and spawn on their return to freshwater but may be driven by instinct to join the spawning migration and subsequently return to the sea as maiden fish.

Larger sea trout include maiden fishes that have spent one or two (rarely more than three) winters at sea, which are returning to spawn for the first time, and proportionately more individuals that have spawned before. Repeat spawning is common among sea trout, exceptionally up to ten times in some individuals, unlike Atlantic salmon, few of which survive to spawn a second time. The run of larger sea trout begins earlier than that of the smaller whitling, usually extending from April to August, but autumn-running sea trout appear to be a feature of some rivers, for example the Tyne and Coquet, and some Welsh rivers. A sea trout of 5kg is a sizeable fish and one of 10kg would count as a specimen fish – a weight achieved by a combination of good growth at sea, longevity and multiple spawning. Sea trout, like Atlantic salmon, are known to home to their natal river.

Habitat

In rivers, juvenile brown trout and sea trout (pre-smolts) require shallow, clean, well-oxygenated, flowing waters with mainly cobble and gravel substrates. Areas with abundant cover, provided by the substrate, undercut banks and aquatic vegetation is favoured. They may also occupy still waters where environmental requirements are similar to those of the river dwelling individuals. Older parr and adult brown trout prefer deeper water, to which fry may also migrate during winter.

Adult brown trout and sea trout both require unimpeded access to their spawning grounds. Sea trout in particular, because of their generally larger size and more extensive movements within the length of a river, also require holding areas to provide refuges on their migration upstream to spawn. They are able to tolerate water temperatures from 0°C up to 30°C, depending on the normal temperature range in which a population has developed, although the limits for growth are in the range 4°–20°C.

Distribution in Britain

The brown trout/sea trout is the only native species of trout in Britain. It is widely distributed, predominantly found in northern and south-western England, Wales, Scotland and the Scottish islands. Brown trout/sea trout are also present in the rivers and chalk streams of the south coast of England. They have a restricted distribution in East Anglia and areas of central and south-east England principally because of unsuitable natural habitat, habitat degradation or poor water quality. Comparison with the distribution mapped in 1972 shows few changes, although recent records imply that brown trout/sea trout are now more widespread than previously in East Anglia and north-east Yorkshire.

World distribution

The brown trout is a native European species – extending northward to Iceland, Scandinavia and Russia and as far south as Spain and into Portugal. To the west it is limited by the European coastline and in the east is found in the Ural Mountains and Caspian Sea. It is also known in the Atlas Mountains of North Africa and in parts of Lebanon.

Sea trout are found along the European coast: north to Scandinavia and the Kola Peninsula; in Iceland; as far east as the Cheshkaya Gulf (east of Archangel); and south to the Bay of Biscay. Although absent from the Mediterranean, populations have been recorded in the Black and Caspian Seas.

Salmo trutta has been introduced to many countries in the last 150 years including Russia, east of the Urals; the USA and Canada; Australia; New Zealand; and several countries in Africa and South America.

Status

- Protected under the Salmon and Freshwater Fisheries Act (1975).
- Intolerant of pollution, affected by habitat degradation, over-abstraction of water and impeded access to spawning grounds.

- Exploitation by rod- and (in the case of sea trout) net fisheries tends to target the larger, more prized individuals and may threaten stocks if excessive.
- The collapse of some sea trout populations on the west coast of Scotland (and in Ireland) appears to be linked to the salmon farming industry – in particular due to lethal infestation of sea trout smolts with sea lice (parasitic copepods) transmitted from farmed salmon.
- Reduced marine survival in Atlantic salmon associated with sea temperature and climate change has not been demonstrated in sea trout.

Hybrids and related species

- Occasionally hybridizes with the salmon *Salmo salar*.
- Related most closely to other species of *Salmo* (salmon) and also to the many species of Salmonidae other than *Salmo* species.

Brown trout, sea trout and Man

Most anglers will, at some time, have caught a brown trout; fewer will have encountered the more prestigious sea trout. However, both brown and sea trout support numerous rod fisheries (and, in the case of the sea trout, net fisheries) of significant economic and social value. In parts of Britain, particularly northern Scotland, brown and sea trout, along with Atlantic salmon, are the only freshwater sport fish. Many populations of brown trout in rivers and still waters have been physically isolated since the end of the last ice age. Most experience little attention from anglers and, in the absence of stocking, retain unique genetic characteristics which reflect their truly wild nature.

In addition to the beauty of these fish, the variety of size and colour they attain and their fighting qualities on rod-and-line, both brown trout and sea trout are valued for their eating – often best appreciated when lightly poached or grilled.

Further reading: Elliott, 1994; Frost & Brown, 1967; Mills, 1971; Solomon, 1994.

Author: Ian Davidson, Environment Agency

Arctic charr *Salvelinus alpinus*

Description

- Medium-sized and streamlined, like a slender trout. Some features vary, including shape of head and mouth.
- Colour very variable, often with light-coloured spots on a grey or brown background. Young and females often drab, but male spawning colours unmistakable bright red underside.
- Small adipose fin.
- Can be confused only with other salmonids, particularly when young with brown/sea trout.

Size

Maximum length typically 35cm and weighing up to 170g.

Biology and behaviour

Spawning usually takes place from October to December, but it has also been recorded from February to March, recognized as so-called 'autumn-' and 'spring-spawning' populations. Such genetically distinct races occur together in each of the two lake basins of Windermere which, when taken together with the stream-spawning population, means that this single water body contains at least five populations of Arctic charr. Spawning is similar to that of other salmonids and involves eggs being deposited and fertilized in a redd constructed by the female within a territory established by the male. Each female lays several hundred eggs, from which young hatch during the spring.

Behavioural interactions with brown/sea trout are an important aspect of the ecology of this species. Studies outside Britain have shown that Arctic charr are usually displaced to an open water habitat when brown/sea trout are abundant inshore. In contrast to the situation at higher latitudes, in Britain Arctic charr also frequently have to co-exist with a variety of other fish species and as a result they typically restrict themselves to offshore habitats outside the spawning season. Among themselves, Arctic charr are more sociable than other salmonids and when in aquaculture they can be successfully kept at very high densities. The diet of Arctic charr is extremely variable and includes zooplankton, particularly cladocerans such as water fleas including *Daphnia* spp., macro-invertebrates including water hoglice *Asellus* spp. and freshwater shrimps such as *Gammarus* spp., and small fishes.

Habitat

Throughout their range in the northern hemisphere many populations of Arctic charr migrate to sea for a short part of their life cycle, but in Britain all are resident in lakes even where easy access to the sea is available. Large, deep, cold, nutrient-poor lakes form the habitat of this species in England and Wales, but in Scotland a significant number of populations also occur in more shallow and productive water bodies. Arctic charr also require a spawning habitat containing a clean gravel substrate, which is typically found in inshore areas or, less usually, in deeper waters or inflowing streams. In Britain, such stream-spawning populations are known in Ennerdale Water, Windermere and Loch Insh, although equivalent populations have been lost from Ullswater.

Distribution in Britain

The majority of British populations of Arctic charr occur in the Scottish Highlands, with much smaller numbers in southern Scotland, the English Lake District and North Wales. Populations have been lost from all three countries and in southern Scotland only one natural population now survives, in Loch Doon, Ayrshire. Eggs, young and adults from this population have been used to establish refuge populations in the Megget and Talla Reservoirs, which are in the catchment of the River Tweed in the Scottish Borders.

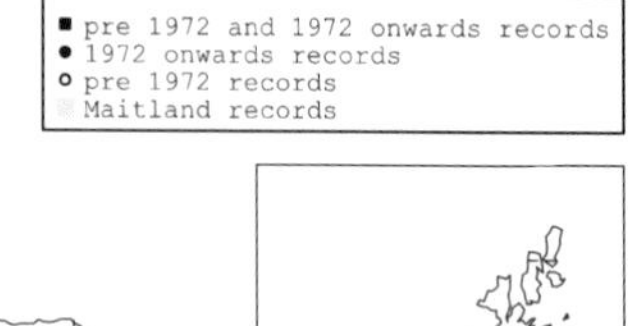

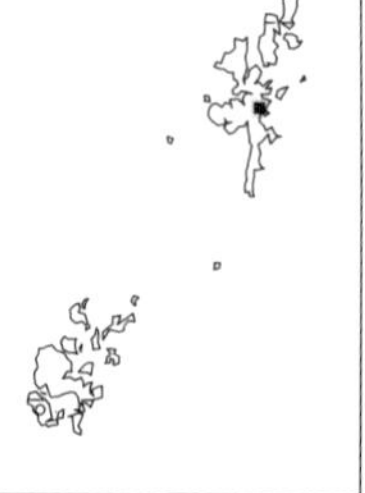

Map 40 Arctic charr

Arctic charr

World distribution

Arctic charr have a holarctic distribution and their ability to undertake migrations to sea has undoubtedly contributed to a near continuous distribution across coastal northern Europe, Asia and North America. In addition, land-locked populations occur throughout this range and form particularly important components of fish communities in many Alpine lakes in Europe.

Status

- Threatened in parts of its European range by factors including eutrophication, acidification, species introductions and over-exploitation.
- Listed under the UK Biodiversity Action Plan, but not a Priority Species.

Hybrids and related species

- No hybrids known in Britain.
- Related most closely to other native game species including Atlantic salmon, trout and grayling.

Arctic charr and Man

The Arctic charr is fished extensively throughout its holarctic distribution where it is taken by gill, trawl and seine nets, traps and various forms of angling including fly-fishing and plumb-lining. In Britain, plumb-lining is still used on Windermere and Coniston Water where small, semi-commercial fisheries supply local restaurants, hotels and fishmongers eager to serve the dark pink flesh of this extremely tasty species. The name *charr*, which should always be spelled with two 'r's, is derived from the Gaelic *ceara* meaning red or blood-coloured.

Further reading: Balon, 1980; Baroudy, 1995; Maitland, 1992.

Author: Ian J Winfield, Centre for Ecology and Hydrology, Lancaster

Brook charr
Salvelinus fontinalis

also known as American brook charr and American brook trout

Description

- Robust but streamlined body with the tail fin squarer compared with other charrs.
- Dark green marbling on back and dorsal fin, red spots with blue halos on sides.
- Underside fins have white edge with inner dark band.
- Mature males more brightly coloured than females, with orange/red flanks and black and white underside.
- Pronounced dark banding on flanks (parr marks) in juveniles.

Size

Up to 40cm long and up to about 1kg in weight, but can be larger for put-and-take fish.

Biology and behaviour

Brook charr spawn in the autumn in streams and areas of upwelling groundwater in lakes, where the eggs are laid and covered in gravel redds. Males compete aggressively at spawning time. The small fry feed on invertebrates and grow

Brook charr or American brook trout

rapidly, with larger fishes taking increasingly larger prey, but only very large brook charr feed on other fishes. Brook charr normally have the shortest life span of all the charrs, especially long-term hatchery strains that grow rapidly to maturity in just over a year and few live longer than three years in areas that are heavily fished by anglers.

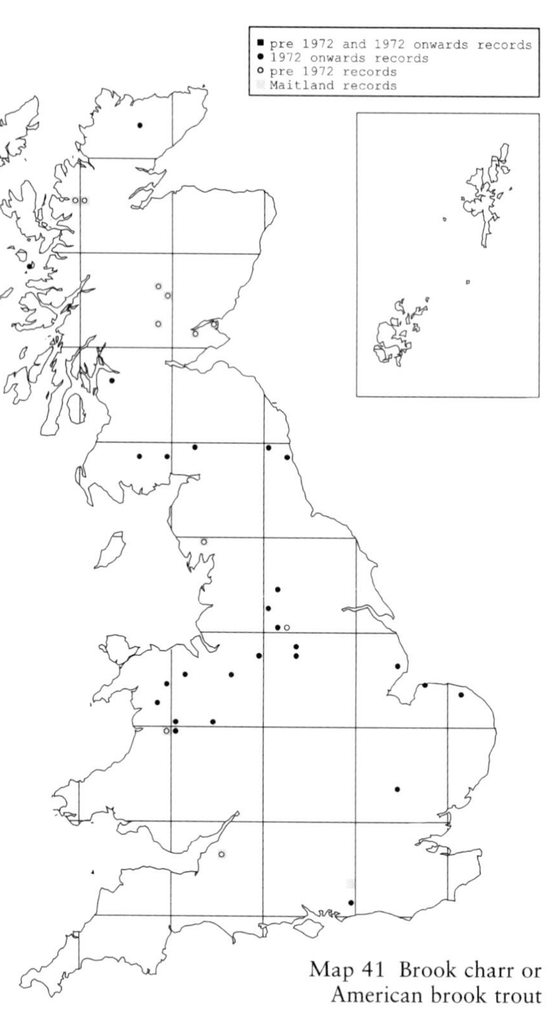

Map 41 Brook charr or American brook trout

Habitat

Brook charr thrive in cool, clear water in well-oxygenated, small to medium-sized rivers and lakes. High summer water temperatures limit their distribution. They are more tolerant of low pH than other salmonids, the lower limit for survival of populations being about pH 4.5. In northern parts of their native geographic range, some fishes known as *salters* migrate to feed at sea close to the river mouths, but this does not occur in Britain.

Distribution in Britain

The species was first brought to Britain in 1868 and has been stocked extensively. Self-sustaining populations occur in only a few locations, nearly all of which were established from introductions made before 1900. In Scotland, long-established and highly-colourful brook charr still occur in the wild including in some small lochs near Lochbuie in Mull (established in 1884), in lochs and associated streams above Torridon, in Wester Ross, and in a small hill loch in the Trossachs. A population in Monzievaird Loch, near Crieff, was established in about 1890 from fishes taken by pony and trap from ponds on the Tayfield Estate, near Newport, Fife, where they have since died out. The population in a small pond and stream on the south side of Loch Tummel, Perthshire, originated in the mid-1970s, but brown/sea trout have been reintroduced there recently and appear to be supplanting the brook charr. In upland areas of Britain, successful spawning occurs sporadically where brook charr are stocked, but the species is seldom able to become self-sustaining in the presence of brown/sea trout. Several commercial put-and-take fisheries offer brook charr, but much less commonly than rainbow trout. The distribution map almost certainly underestimates the extent to which this species has been introduced in Britain.

World distribution

Native to streams and lakes in cool to temperate regions of eastern North America. It is naturalized in continental Europe, Asia, New Zealand and Australia, southern Africa and South America.

Status

- All species of the genus *Salvelinus* (excluding the native *Salvelinus alpinus*) are listed on the Prohibition of Keeping or Release of Live Fish Orders (see Appendix 3) as species for which release to the wild is not permitted without a licence.

Hybrids and related species

- Closely related to and hybridizes readily with Arctic charr and lake charr *Salvelinus namaycush* (Walbaum), producing fertile offspring, though rarely in the wild.
- Hybridization with brown/sea trout and Atlantic salmon occurs less readily; the offspring are infertile.

Brook charr and Man

Brook charr are beautiful fish that are delicious to eat, the flesh often developing a deep orange colour. Farmed strains are relatively easy to cultivate and they grow fast, but brook charr are much less popular than rainbow trout for aquaculture, perhaps because of their preference for cooler water. Brook charr are regarded as one of the easiest of the salmonids to catch by angling, and in Britain they are stocked to add variety to some intensive angling fisheries.

Further reading: MacCrimmon & Campbell, 1969.

Author: Andy Walker, FRS Freshwater Laboratory

European grayling *Thymallus thymallus*

Description

- Streamlined, slightly flat-sided, silvery grey-green back and flanks with black flecks, whitish under-body.
- Very large dorsal fin with four or five rows of red and black spots. Dorsal fin of male is larger than that of female. Small, dark adipose fin and narrow 'wrist'.
- Mouth relatively small with upper jaw protruding beyond lower jaw. Large eye with pear-shaped pupil.
- Skin and flesh have faint smell of the herb thyme.
- Juveniles silvery/light green with bluish parr marks along flanks.

Size

Average length about 30cm, but can grow to over 50cm and over 1kg in weight. A 1kg grayling is regarded as a specimen fish.

Biology and behaviour

As winter approaches, grayling tend to form shoals. Spawning starts when water temperatures reach about 4°C, between late February in the south and April in the north, but usually in March and April. Males defend territories in clean, moderately flowing water with a gravel substrate, and court females approaching them from downstream. With a vibrating display and erect fins, the male initiates spawning, which generally occurs in the afternoon or evening. The fertilized eggs are deposited in a redd near the gravel substratum, and the spawning act can result in burial of eggs to a depth of 5cm. Females are promiscuous and can spawn with many males.

European grayling

Juvenile grayling grow quickly, reaching approximately double the weight of other Salmonidae (trout and salmon) in the first year. In southern waters grayling mature after two to three years and most die by their fifth or sixth year. In cooler, more northerly rivers they mature and grow more slowly and therefore live proportionately longer. Growth slows markedly with the onset of sexual maturity.

Grayling feed throughout the day and rest near the substrate at night. They feed opportunistically on invertebrates in the water column and near the substrate but will also take terrestrial invertebrates from the surface. Their prey varies with age, with older fishes tending to feed lower in the water column, near to the bottom.

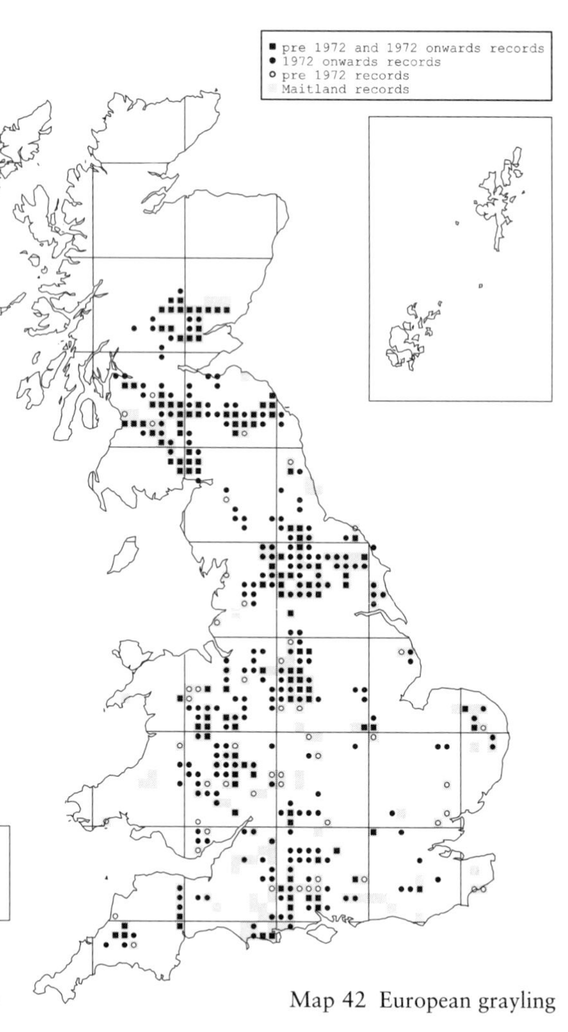

Map 42 European grayling

Habitat

The slow-flowing marginal zones containing vegetation provide vital nursery areas for grayling larvae and fry. Larger fry occupy the transition area between the margins and mid-channel with a sand and pebble substrate where they catch drifting prey. Adult grayling favour clean, well-oxygenated, faster flowing water with a pebble substrate during summer, and deeper, slower flowing water in winter. At night they move to shallower water.

Grayling are unable to tolerate high water temperatures (18°–25°C) without acclimation and are very susceptible to organic pollution. In rivers where their habitat requirements are fulfilled, the home range may not extend beyond a few hundred metres but, in some large rivers in continental Europe, grayling undertake large migrations extending beyond 100km to feeding and spawning grounds.

Distribution in Britain

European grayling are thought to be native to parts of England and possibly Wales, having reached Britain before its isolation from continental Europe after the last glacial retreat. However, all Scottish populations of grayling, and those in many other British rivers, are thought to result from introductions within the last 150 years. They are widely distributed throughout Britain and found in most river types from northern spate rivers to southern chalk streams. Their distribution is restricted in eastern England due to unsuitable natural habitat, habitat degradation and obstructions to their natural passage in rivers. Grayling are also known from a few still waters, including the Gouthwaite reservoir on the River Nidd in Yorkshire and Llyn Tegid in North Wales. Comparison with the distribution in 1972 shows a broadly similar pattern, with some apparent losses and few gains.

World distribution

European grayling occur in clean rivers and lakes throughout northern and eastern Europe, from Wales to the White Sea and upper tributaries of the Volga, but many populations are considered under threat. Although in Britain grayling spend their life in fresh water, elsewhere they can occasionally occur in brackish water.

Status

- Exploitation is regulated under the Habitats and Species Directive and the Bern Convention and listed under the UK Biodiversity Action Plan.
- Threatened throughout its range by over-fishing, pollution, and physical changes to the habitat which restrict access to spawning sites.

Hybrids and related species

- Occasionally hybridizes with Arctic grayling *Thymallus arcticus* (Pallas) where their ranges overlap in Russia.
- Most closely related to other *Thymallus* species and subspecies, variously referred to as the Amur, Baykal, Hovsgol and Mongolian graylings.

Grayling and Man

Grayling were previously regarded as vermin, particularly in southern chalk streams where anglers favoured brown trout. Consequently they were systematically removed in large numbers from trout streams. Attitudes toward this species have changed and grayling are now more valued as a fishery resource. Fishermen appreciate the fine sporting qualities that the 'Lady of the Stream' offers during the winter months, outside the trout season. The flesh is palatable with a taste preferred by some to that of trout.

Further reading: Broughton, 2000.

Author: Richard Cove, Environment Agency

Burbot *Lota lota*

Description

- The only freshwater member of the cod family in Europe; strongly resembles the marine ling *Molva molva* (Linnaeus).
- Slender, smooth skinned, with single large barbel in middle of chin.
- Colour overall yellow-olive to olive-green with dark marbling on sides and back, and under-side off-white to yellow.
- First dorsal fin short and second very long, reaching almost to tail.

Size

Normally adults grow to 30 cm in length and around 3kg in weight.

Biology and behaviour

Burbot spawn in mid-winter or very early spring at temperatures of 0.5°–4°C. They spawn at night in small groups in shallow water, but in lakes they can spawn in deeper areas. A group of around twenty males and females form a writhing ball and release eggs and sperm. The fertilized eggs are very small and a female can produce 500,000 eggs per kilogram of body weight. Eggs hatch after 35 to 70 days depending on tempera-ture, and the fry appear in the spring around five weeks after fertilization. Burbot fry feed on rotifers and copepods, as do juveniles, which also feed on small molluscs. Adults feed on insects and small fishes, as well as fish eggs. If available, perch and roach are a preferred food item.

Habitat

Burbot occur in a wide range of water types: from large lakes, for example the Great Lakes of

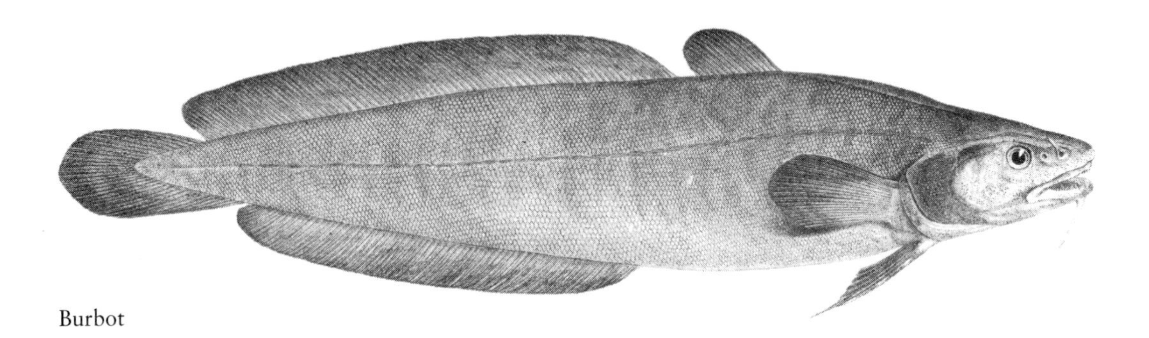

Burbot

Canada, to cold or snow-melt rivers, such as the Vltava in the Czech Republic or the Rhône in France. They appear to prefer habitats with complex physical features such as tree roots, rock piles, or submerged logs.

Distribution in Britain

Burbot were formerly found in rivers flowing into the North Sea that would have been connected to the Rhine in the early post-glacial period. Their range extended from Durham to East Anglia, including the Yorkshire Derwent, the Trent, the Cam and the Nene. The possible causes for extinction include pollution, warming of water, especially in winter, and the break-up of spawning migrations due to land drainage practices.

World distribution

It has a circumpolar range from northern Europe south to Romania and France, and extending through Russia and Asia to North America and Canada.

Status

- Extinct in Britain, last confirmed in Cambridge in 1969
- Listed under the UK Biodiversity Action Plan.
- Listed on the Prohibition of Keeping or Release of Live Fish Orders (see Appendix 3) as a species for which release to the wild is not permitted without a licence.

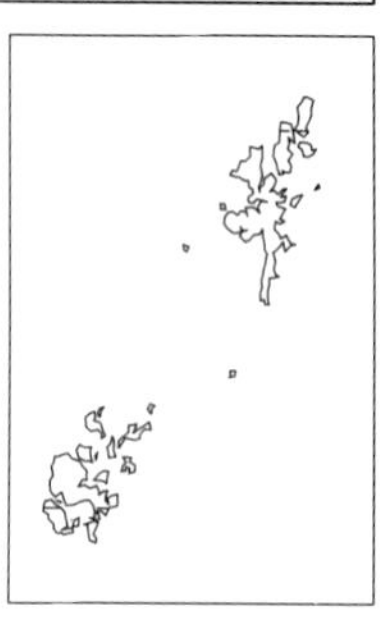

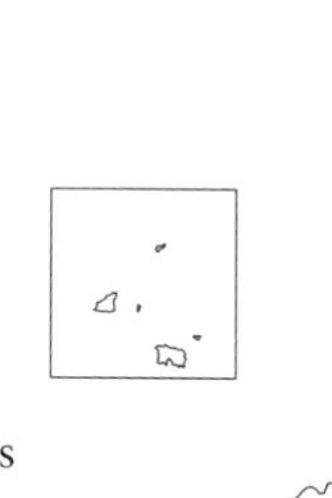

Map 43 Burbot

Hybrids and related species

- None in fresh water.

Burbot and Man

Burbot has many names throughout the world, the most common English ones being poult and eel pout. It is considered very good eating by knowledgeable fisherman and when boiled and buttered it has been called ‘poor man’s lobster’. The liver is a special delicacy and is used in *pâtés* and Lenten garnishes. A French country proverb suggests that a housewife would sell her soul for a burbot liver!

Further reading: Marlborough, 1970.

Author: Keith Easton, Environment Agency

Three-spined stickleback
Gasterosteus aculeatus

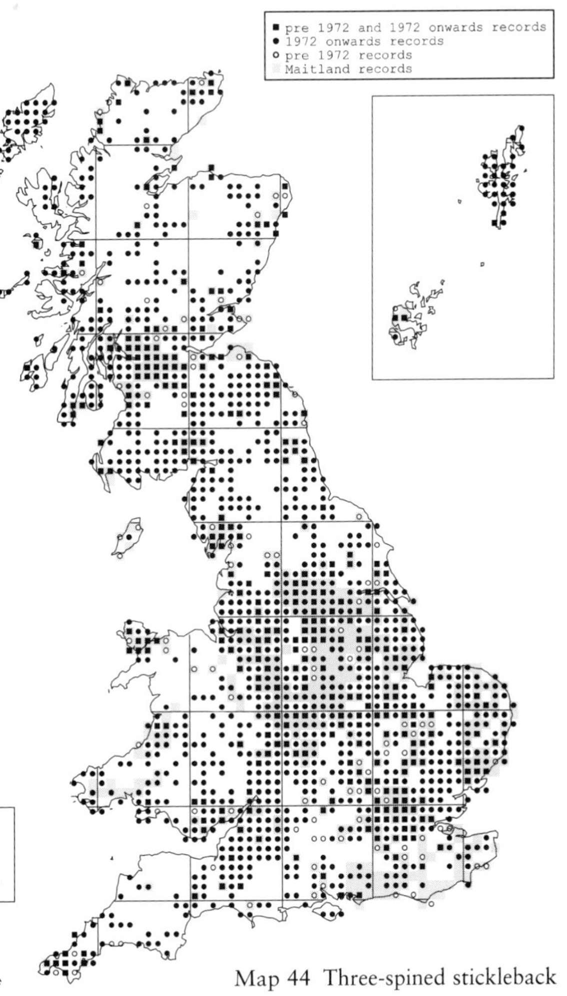

Map 44 Three-spined stickleback

Description

- Small with elongate, laterally compressed body tapering to slender 'wrist'.
- Three or four spines aligned in front of dorsal fin; pelvic fins each with prominent spine. Spines can be locked rigidly erect. Forms without spines occur but are rare.
- Scaleless but with pelvic girdle of bony plates and some lateral bony plates.
- Olive to dark green above, sometimes with darker mottles, underside silvery. Fins pale, membranes may be orange/red.
- Gill covers do not meet ventrally.
- Breeding males develop vivid red belly and flanks with bright blue iris.

Size

Normally about 5cm; maximum size in fresh water approximately 8cm.

Biology and behaviour

Three-spined stickleback are short-lived, usually surviving to breed only once. In southern England, the typical life span is about 16 to 18 months, but further north populations over three years old are known. Three-spined sticklebacks migrate to breeding areas in spring, stimulated by increasing day length and water temperatures, and the peak of spawning is in late spring/early summer. Populations resident in fresh water migrate to shallows and backwaters, and coastal fishes move inland. The spawning behaviour is highly ritualized. Each breeding male defends a territory and constructs a nest on the bed of the watercourse made of plant fragments and other detritus, glued together with 'spiggin', an adhesive protein secreted from the kidneys. Ripe females are encouraged into the nest to deposit their eggs, which are then fertilized by the male. The male then drives off the female and cares for the nest alone, 'fanning' oxygenated water over the nest for ten days or more until hatching, defending the nest and its contents against predators (including cannibalism by other sticklebacks) and retrieving fry for the first few days after hatching. Eggs from several females may be deposited in one nest and a single male may have several nests during the course of a breeding season. Both sexes can remain in breeding condition for several weeks. The developing fry and any surviving adults make the return migration in autumn/winter.

Three-spined sticklebacks are omnivorous, depending on their size and what is available in

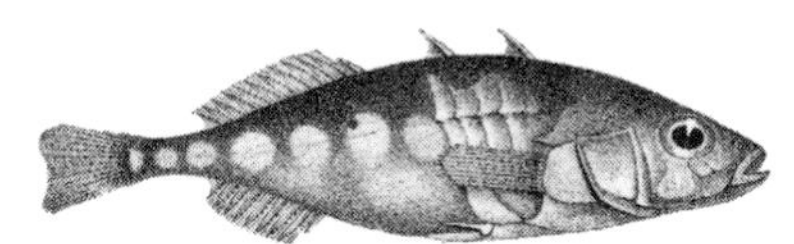

Three-spined stickleback

the habitat. Food items include small invertebrates, especially worms, crustaceans and aquatic insects, as well as the eggs and fry of fishes.
The internal parasites of sticklebacks have been studied in some detail (see Further reading) and, in the wild, any individual three-spined stickleback may be host to a range of over 70 species of parasite.

Habitat

Three-spined sticklebacks normally spawn among sandy or silty sediment in slow-moving shallow, fresh or low-salinity water. Shoaling fry can be observed in open water, while older fish are often associated with aquatic plants. They appear more tolerant of physical habitat degradation and pollution than do other species of freshwater fish in Britain.

Distribution in Britain

Three-spined sticklebacks are found in all types of fresh waters throughout Britain and are probably absent only from high altitude lakes and rivers. They are, however, often absent from river reaches at lower altitudes that have swift currents and few refuges. The apparent expansion in range since 1972 is almost certainly a result of increased recording in the intervening period, and this species is probably still under-recorded in some areas.

World distribution

Almost circumboreal, absent only from parts of the Arctic coasts of northern Siberia and North America. Found along both the Pacific and Atlantic coasts of North America and throughout the Atlantic provinces of Canada. Occurs in southern Greenland, Iceland, Ireland and Britain, and in continental Europe from northern Norway around the coastline (including the Baltic), through the northern coast of the Mediterranean to the Black Sea. Common in estuaries, coastal waters and in the surface waters of the high seas.

Status

- Not threatened in Britain, the rest of Europe or globally.

Hybrids and related species

- Hybrids with nine-spined stickleback, its only related species in British fresh waters, have been bred under artificial conditions, but in the wild the two species are effectively isolated from each other by their reproductive behaviour.
- Unmistakable when fully grown, but juveniles may be confused with those of the nine-spined stickleback.
- Several morphs of three-spined stickleback have been described, based on the number of lateral bony plates present. These seem to increase in number in populations living in higher levels of salinity or at sea.
- Fifteen-spined or sea stickleback *Spinachia spinachia* (Linnaeus) is classed as a wholly marine species and is common in coastal waters.

Three-spined sticklebacks and Man

Sticklebacks are probably the most familiar wild fishes in Britain, being the first target of many pond dippers. The three-spined stickleback is one of the better known vertebrates; because it is common, small, easy to capture and robust, it makes an ideal subject for aquarists, natural historians and scientists alike. Although of no current commercial interest (other than the occasional appearance among 'whitebait'), three-spined sticklebacks have been harvested in the past for their oil and for use as agricultural fertilizer. Their importance as food for predatory fish species, such as pike and trout, tends to be overlooked in relation to angling revenues.
The three-spined stickleback is probably unique among the fish fauna of Britain in being sampled from water with a shotgun as part of a scientific research project!

Further reading: Arme & Owen, 1964; Coad, 1981; Craig-Bennett, 1931; Donoghue, 1988; Elkan, 1962; Mann, 1971; O'Hara & Penczak, 1987; Ramage, 1825; Scott & Crossman, 1973; Vik, 1954; Wootton, 1976, 1984.

Author: Willie Yeomans, Clyde River Foundation

Nine-spined stickleback
Pungitius pungitius

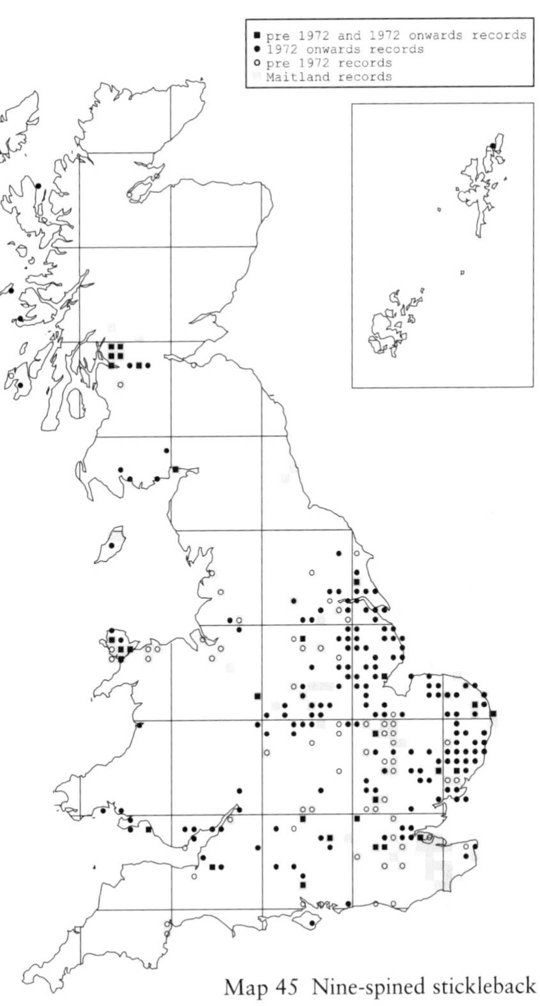

Map 45 Nine-spined stickleback

Description

- Small with elongate, laterally compressed body tapering to very slender 'wrist'.
- More than six (usually nine or ten) spines in front of dorsal fin, aligned to alternate sides, forming V or W shape when viewed from front; pelvic fins each with prominent spine. Spines can be locked rigidly erect.
- Scaleless but with pelvic girdle of bony plates and sometimes with a few small lateral bony plates.
- Olive to dark green, light brown or golden above, undersides silvery. Fins pale.
- Gill covers merge ventrally.
- Breeding males develop black belly and flanks, with pelvic spines contrasting white.
- Juveniles may be confused with those of three-spined stickleback, but unmistakable when fully grown; obviously more slender and delicate-looking than three-spined.

Size

Smallest British freshwater fish, rarely more than 6cm in length.

Biology and behaviour

There is little information on the lifespan of the nine-spined stickleback in Britain; a population from the Wirral, Cheshire, is known to live for more than three years, but this may not be typical. Spawning occurs from spring to mid-summer, stimulated by increasing day length and temperatures. Each male holds a territory in which a nest is constructed of plant material glued together with a secretion from the kidneys. The nest has a tunnel through it and is usually suspended between the stems of waterplants. Several females are encouraged to deposit eggs in each nest, which are then fertilized by the male.

Nine-spined stickleback

The male cares for the nest alone. The eggs hatch after six to ten days, and the male may construct a 'nursery' for the hatchlings from vegetation above the nest. The fry disperse individually two weeks after hatching.

Nine-spined sticklebacks are visual predators which feed mainly on small invertebrates, particularly planktonic crustaceans and midge larvae, but they will also cannibalize stickleback eggs and larvae. Some fifty species of parasites are associated with nine-spined sticklebacks (see Further reading).

Habitat

The habitat preferences of nine-spined stickle-backs are generally similar to those of the three-spined, but where the two species occur together, the preference of the nine-spined for plant cover is very noticeable. Spawning occurs within weed beds or occasionally on the bottom in fresh water and nine-spined sticklebacks may be intimately associated with plants, sometimes in very shallow water, throughout their lives. They can be abundant in favourable habitats, and appear particularly well suited to life in weedy ditches where they are better able to tolerate low levels of dissolved oxygen than three-spined sticklebacks.

Distribution in Britain

Nine-spined sticklebacks live mainly in the shallows of slow-flowing fresh waters in England, Wales and parts of southern and western Scotland. They can also be found in coastal waters, but not on the high seas around Britain. Comparison of the current known distribution with that up to 1972 suggests an increase in range in some areas but a decrease in others, but this is almost certainly a false picture. It may result from the overall small amount of sampling that has occurred because the species is of little importance in assessing water quality or in maintaining fisheries and is of no commercial value.

World distribution

Although truly circumboreal, the range of nine-spined sticklebacks in inland waters and in coastal waters varies considerably. In North America it does not extend south of Alaska on the Pacific coast nor south of New Jersey on the east coast, but does penetrate inland to the Great Lakes. In Europe, although absent from Iceland, it is distributed from the Arctic Ocean, throughout Scandinavia and the Baltic, to the Atlantic coast of France and in Ireland. In mainland Asia and Japan its distribution is entirely coastal.

Status

Not threatened in Britain, the rest of Europe or globally.

Hybrids and related species

- Only closely related species in British fresh waters is the three-spined stickleback, with which hybrids have been bred under artificial conditions, but no hybrids have been found in the wild.
- Fifteen-spined stickleback *Spinachia spinachia* (Linnaeus) is wholly marine and is common in coastal waters.

Nine-spined sticklebacks and Man

This somewhat cryptic fish is poorly known and little studied, although it provides forage for several species of commercial value.

Further reading: Coad, 1981; Dartnall, 1973; Scott & Crossman, 1973; Wootton, 1976, 1984.

Author: Willie Yeomans, Clyde River Foundation

Bullhead *Cottus gobio*

Description

- Small, bottom-dwelling, with broad, flattened head and tapering body.
- Large eyes set high on head.
- Smooth, unscaled skin, mottled brown but pigmentation may vary according to background.
- Opercular spine and spiny rays in first dorsal fin and well-developed pectoral fins afford protection from predators.

Size

Maximum length usually 8 to 10cm.

Biology and behaviour

Bullheads are usually solitary and during the summer and winter months males become aggressively territorial, defending a home range to which they show strong homing instincts if displaced.

During the spawning period (March to April), the male excavates a nest under stones, into which females lay clusters of eggs, usually on the ceiling, before leaving the male to guard and care for them until they hatch some three to four weeks later. The male fans water with his pectoral fins over the eggs to keep them well oxygenated. As he is unable to forage at this time the male often loses condition and in extreme circumstances this may lead to the cannibalism of eggs from his own nest. Desmond Morris observed that a male guarding his nest has one simple response to anything that moves near the nest entrance: he bites it. If the item is food, it is swallowed; if it is a pest, it is spat out; if it is a female, it is spat into the nest; and if it is a rival, it is carried away from the entrance. Upon hatching, juveniles move away from the nest to avoid being eaten by the male who once more takes up a solitary existence until the following breeding season.

Bullheads feed mainly on benthic fauna such as copepods and other crustaceans during the winter months, and the nymphs and larvae of aquatic insects such as mayflies and caddisflies during the summer. There is little evidence to support the frequent accusation that bullheads prey upon the eggs and juveniles of salmonid species.

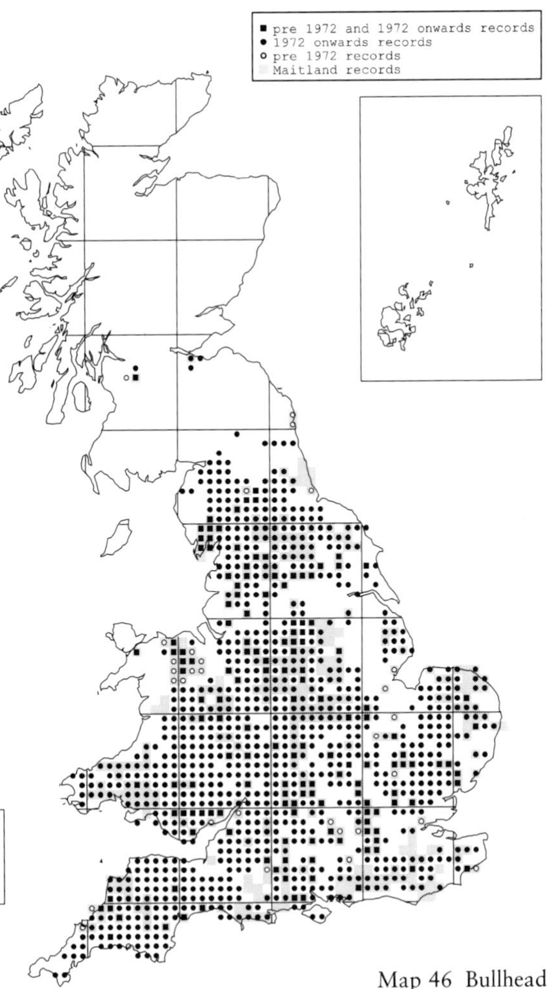

Map 46 Bullhead

Habitat

Bullheads are vulnerable to predation by piscivorous fishes such as trout, and birds such as kingfishers and herons. They normally seek secure refuges during the day emerging at dawn and dusk to forage. Adults usually hide among the interstitial spaces of large gravel, woody debris or leaf litter, preferring this to the partial cover provided by weed or small gravel. Juveniles are often excluded from these preferred refuges by the larger adults and seek safety in more exposed sites such as stony riffles. As they grow, young fishes locate better sites offering more cover and protection.

Distribution in Britain

Bullheads were thought to be indigenous only to the south-east of England, but to have spread within lakes and rivers throughout England and Wales. This view has been challenged by recent analyses of the mitochondrial DNA of bullhead populations in Europe, which suggest that populations persisted within the south-west of England throughout the last major glaciation. This also challenges the previously held belief that British freshwater fish assemblages consist exclusively of post-glacial colonists. The current distribution of bullheads may reflect both the natural, long-term dynamic processes of river catchments, as well as introductions and translocations by humans. Although the habitat and temperature of many rivers and lakes in Scotland are suitable for this

Bullhead

species, bullheads are scarce in the region, having been recorded in the Clyde catchment (White Cart Water) and southern tributaries of the Firth of Forth.

World distribution

The bullhead is widely distributed in streams and lakes throughout central Europe. Populations extend from the Pyrenees through the North European Plain to the foot of the Ural Mountains in Russia. They are absent from Ireland.

The bullhead is a member of the sculpin family, Cottidae, which are found within the seas and freshwater systems of Europe, Asia and North America. Generally small, sedentary, bottom-dwelling fishes, they feed rapaciously on crustaceans and other small fishes.

Status

- Protected under the Habitats and Species Directive and listed under the UK Biodiversity Action Plan.

Hybrids and related species

- No hybrids known in Britain.
- The sculpin family contains over 300 species, most of which are marine. Three or four species occur in European fresh waters, but the bullhead is the only species found in Britain. The alpine bullhead *Cottus poecilopus* Heckel occurs in Scandinavia and central Europe through to Asiatic Russia, and the fourhorn sculpin *Triglopsis quadricornis* (Linnaeus) inhabits brackish waters on arctic shores and the depths of isolated, large lakes. A fourth species, *Cottus petiti* Bacescu & Bacescu-Mester, is recognized from the catchment of the river Lez in southern France, but it may be only a subspecies of *C. gobio*.

Bullheads and Man

The bullhead, also known as the miller's thumb, has little economic value or interest for the angler, other than as live-bait, or as an aquarium species, an environment in which they can soon become tame. Formerly, the sweet-tasting flesh of the bullhead seemed to be valued, and Isaak Walton and some of the older authorities are enthusiastic about it as a food. The British rod-caught record stands at 28g, caught in the Green River, Guildford, Surrey, in 1983.

Further reading: Hänfling *et al.*, 2002; Marconato *et al.*, 1993; Morris, 1954; Mills & Mann, 1983; Perrow *et al.*, 1997; Volckaert *et al.*, 2002.

Author: Richard Horsfield, Environment Agency

Pumpkinseed *Lepomis gibbosus*

Description

- Unlike any other freshwater fishes found in the wild in northern Europe. Deep, laterally compressed body with large scales. Head length much less than body depth.
- Dorsal fins continuous with 10–11 strong spines.
- Back and upper sides golden brown to olive, lower sides golden with irregular wavy blue-green lines, underside bronze to orange red. Dusky bars on sides. Opercula with wide black spot.

Size

Length 10–15cm; weight under 100g.

Biology and behaviour

Pumpkinseed spawn between May and October when water temperatures are at least 20°C. Males are territorial and guard the nest until the young hatch. These mature at two to three years, feeding on small invertebrates, fish eggs and young fishes.

Habitat

Shallow, sheltered, weedy still waters or slow-flowing rivers.

Distribution in Britain

Introduced from North America early in the twentieth century, they are probably restricted to southern England and are known to breed here. Anecdotal evidence suggests that some populations may have been established from specimens that had escaped or been released from research facilities.

World distribution

Native to the warmer regions of north-eastern North America, they were introduced to Europe at the end of the nineteenth century as ornamental species. They have become well established in areas from southern France through to the Baltic States, southern Russia and the Danube Delta.

Status

- Listed under the Wildlife and Countryside Act 1981 and on the Prohibition of Keeping or Release of Live Fish Orders (see Appendix 3) as a species for which release to the wild is not permitted without a licence.

Hybrid and related species

- Closely related to the red-breasted sunfish *L. auritus* (Linnaeus) and the green sunfish *L. cyanellus* Rafinesque, both of which are now found in Europe.

Pumpkinseed and Man

Their coloration makes them attractive aquarium and ornamental pond species. In North America, where they reach larger sizes, they are harvested commercially for food.

Further reading: Lever, 1977.

Authors: Sarah Chare & Robin Musk, Environment Agency

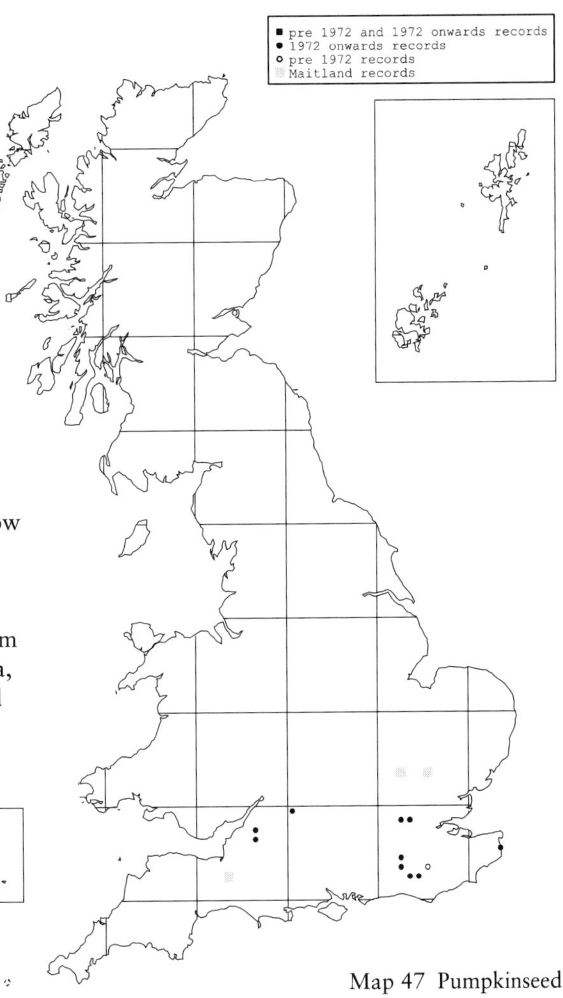

Map 47 Pumpkinseed

Pumpkinseed

Largemouth bass
Micropterus salmoides

Description

- Body shaped like an elongated perch; dorsal fin deeply notched with anterior half spiny.
- Upper head and back dark olive, sides paler and underside whitish.
- Mouth very large with protruding lower jaw.

Size

Adults can grow to almost 1m in length and weigh up to 11kg.

Biology and behaviour

The largemouth bass spawns annually from May to August, once water temperature reaches 16°C, laying eggs in a nest dug into sand or gravel. Males guard their nests and each brood stays together for several weeks after hatching. To capture prey, bass lie in ambush and dash upwards towards the surface. Juveniles feed on aquatic invertebrates and small fishes; adults also eat fishes, as well as crayfish and frogs.

Habitat

The largemouth bass inhabits clear, vegetated lakes and the backwaters of rivers, preferring quiet water and overgrown banks. It tends to occupy mid-water, living in depths to seven metres, within a temperature range of 10°–32°C.

Map 48 Largemouth bass

Largemouth bass

Distribution in Britain

It is believed that all British populations of largemouth bass have now died out. The earliest introduction appears to have been to Argyll, Scotland, in 1881. Further introductions, mostly unsuccessful, were made in England during the mid-1920s and the 1930s, including sites in Cornwall, Surrey, Kent, the Midlands and Manchester. By the mid-1970s it was suspected that the only surviving population was in a disused clay-pit near Wareham, Dorset. It is likely that low temperatures during the breeding season are the main reason for failure of the species to become established in waters that otherwise appear suitable.

World distribution

The largemouth bass occurs naturally in the eastern and southern USA and northern Mexico. The species has been successfully introduced elsewhere as a sport fish and has attained an almost worldwide distribution. In some countries, such as Spain, it has been blamed for the local disappearance of native species of fishes.

Status

- Listed under the Wildlife and Countryside Act 1981 and on the Prohibition of Keeping or Release of Live Fish Orders (see Appendix 3) as a species for which release to the wild is not permitted without a licence.

Hybrids and related species

- No hybrids known.
- One of six species of freshwater bass worldwide, and in the same family as the pumpkinseed.

Largemouth bass and Man

The largemouth bass is one of the most popular sport fishes of North America. It responds well to lures of all kinds and is highly regarded by anglers. It provides good eating when skinned, filleted and fried in batter, the meat being white, flaky and low in oil content.

Further reading: Lever, 1977; Wintle, 2001.

Author: Phil Hickley, Environment Agency

Ruffe *Gymnocephalus cernuus*

Description

- Small member of the perch family with large eyes and mouth.
- Dorsal fins joined, spiny at front and soft at rear.
- Back and flanks sandy to pale, greenish brown with irregular darker blotches but no distinct stripes. Underside pale yellow or silver.
- Scale edges finely toothed, giving rough body surface.
- Gill cover ends in large spine.

Size

Length seldom more than 15cm; exceptionally weighs up to *c.*150g.

Biology and behaviour

Ruffe normally live in shoals. In larger waters, both rivers and lakes, mature fishes move into relatively shallow water to spawn, usually in water up to a few metres deep over a sandy or stony bottom. Mature females usually produce two batches of eggs each year, during May and June. The eggs, which are sticky and pale yellow, are shed in strands and hatch after about 10 days. The newly hatched fry are about 3cm in length. Initial growth is rapid, and fry grow to 5–6cm by the end of the first summer. Thereafter, growth is slower and after two summers they are usually 8–10cm in length. In Britain, few ruffe live longer than four years; females generally grow faster and live longer than males. Many males and some females spawn at a year old and all are mature at two years old.

In rivers, insect larvae, including chironomids and caddis, and crustaceans, particularly ostracods, copepods and water hoglice, are major food items. In some lakes and reservoirs, planktonic crustaceans, including cladocera, are important prey. In some waters, predation of the eggs of other fishes has been observed, including those of coregonids, and the impact of such predation on uncommon species has recently been a topic of considerable interest.

Habitat

Ruffe occur in a wide range of freshwater habitats but are absent from small ponds and fast-flowing rivers. In rivers, high densities can occur over sandy and stony substrates, sometimes in association with shoals of gudgeon. In lakes, ruffe can be found at considerable depths, provided oxygen levels are high enough.

Distribution in Britain

Formerly confined to the catchments of English lowland rivers draining into the North Sea, this species has spread slowly westwards and northwards. Since deliberate introductions of ruffe are unlikely, most of this range expansion is probably due to pike anglers discarding unused livebait and to inadvertent introductions of ruffe during stocking of other coarse fish. Since 1972, ruffe have become established in Llyn Tegid (Bala Lake), Loch Lomond and in the Lake District, notably Bassenthwaite Lake. In most waters, ruffe are a minor component of the fish community but, in disturbed or newly colonized waters, they may constitute a considerable proportion of the total fish stock and jeopardize populations of native fishes (see whitefish, p. 106). During the 1970s, many ruffe populations appeared to suffer mortalities due to 'perch ulcer disease' and many populations may not have recovered to their former levels. Long-established populations of ruffe often have a variety of parasites. Infestations of some, such as the larval fluke *Ichthyocotylurus platycephalus* (Creplin), found mainly in the body cavity, may be so heavy that they increase mortality and reduce stocks. It is likely that the rapid expansion of ruffe populations in newly colonized waters may be aided by initially lower infestations of parasites.

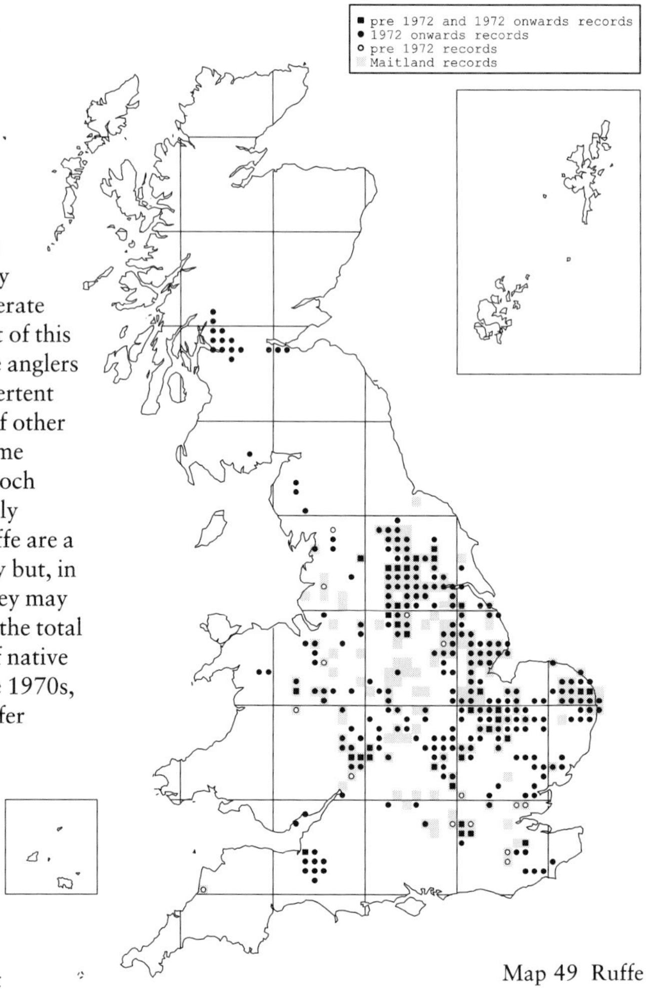

Map 49 Ruffe

Ruffe

World distribution

Ruffe are found in many fresh waters in northern Europe and central and northern Asia. They also occur in areas of the Baltic Sea with low salinity, where growth is faster and maximum size is greater than in fresh waters. Notable range expansions have been to Bodensee (Germany/Switzerland), Norway and Italy. Their introduction to the Great Lakes of North America appears to have occurred through transport in ballast water of a ship.

Status

- Not threatened in Britain or elsewhere.
- Listed on the Prohibition of Keeping or Release of Live Fish Order (Scotland) (see Appendix 3) as a species for which release to the wild is not permitted without a licence.

Hybrids and related species

- No hybrids known in Britain.
- There are three other species of ruffe, Don ruffe *G. acerina* (Güldenstädt), Balon's or Danube ruffe *G. baloni* Holcík & Hensel, and schraetzer or striped ruffe *G. schraetser* (Linnaeus), which have restricted distributions in eastern Europe.

Ruffe and Man

Exploitation of ruffe has been limited and they are rarely deliberately targeted by anglers, but may be captured incidentally when fishing for other species. Ruffe are used as live or dead bait for predatory species such as pike, and in the past have been used for human consumption. In some waters, especially those that have been colonized recently, ruffe may be considered undesirable. In particular, they are regarded as a competitor for food with bream and perch and they feed on the eggs of rare coregonids, such as vendace and whitefish. Why the ruffe, whose name is perhaps suggested by the roughness of its scales, has so many other vernacular names in Britain is unclear; they include jack-ruffe, pope, tommy, cocky and daddie.

Further reading: Gunderson, *et al.*, 1998; Ogle, 1998; Winfield, *et al.*, 1998.

Author: David Hopkins, Environment Agency

Perch *Perca fluviatilis*

Description

- Distinctive deep body with normally four to six or sometimes up to nine dark vertical bars over olive green flanks. Sometimes silver, pale blue or grey specimens occur.
- Belly usually cream, occasionally orange.
- Two dorsal fins. The most prominent is opaque and heavily armoured with sharp spines. Pelvic, anal and caudal fins are typically orange to deep red.
- Head heavily armoured with bony plates. Large mouth with numerous small teeth.

Size

Adults range from 20–40cm in length and 1–2.5kg in weight.

Biology and behaviour

Perch can live for up to ten years in productive waters, where males mature after two years, and females after four years. Spawning takes place during the early spring, when water temperatures reach approximately 10°C. Adult female perch then actively seek shallow areas with dense submerged vegetation before shedding a series of long, gelatinous ribbons containing up to 200,000 eggs. Perch are predatory and, from an early age, will readily eat other small fishes. They even cannibalize each other, which enhances a population's chance of survival when there is a limited supply of food. They also feed on aquatic invertebrates and on terrestrial insects that have fallen into the water.

Habitat

Many anglers often perceive perch as living in deep, dark haunts away from fast flows and within striking distance of their prey, such as fry. This is partly true, but perch are gregarious, shoaling fishes that are to be found in open water as well as around complex underwater structures. Perch will use submerged or emergent vegetation as places of ambush, as well as open water that provides little or no cover for their prey. Thus, perch utilize habitats that enable them to capture their prey most effectively, whilst also evading capture themselves. Perch are adapted to living in both still water and fast-flowing rivers, at a variety of depths, ranging from only a few centimetres to 30m.

Distribution in Britain

Perch are now widespread throughout England, Wales and Scotland, including some remote waters in upland areas, reflecting their ability to survive in a multitude of conditions and within a range of quite extreme habitat types. This wide dispersal to remote areas and even to islands, such as Mull and the Isle of Man, has probably resulted from unrecorded introductions by humans. Anecdotes based on folklore claim that the unusual and often seemingly remote occurrences of perch are due to the transportation of ribbons of eggs by water birds to areas outside the fish's natural range.

World distribution

Perch are distributed throughout most of western, northern and central Europe and into Asia. They have been introduced to the Iberian Peninsula.

Status

- Not threatened in Britain or continental Europe.
- One of the most widespread freshwater fish species in Europe.

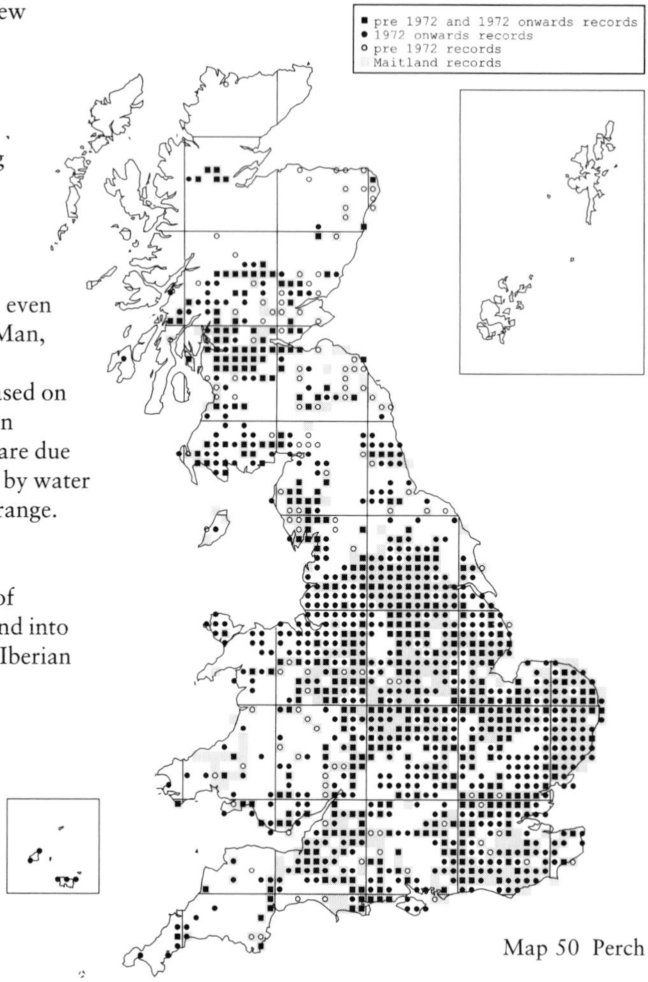

■ pre 1972 and 1972 onwards records
● 1972 onwards records
○ pre 1972 records
Maitland records

Map 50 Perch

Perch

Hybrids and related species

- No hybrids of perch are known.
- The only other *Perca* species is the American yellow perch *Perca flavescens* (Mitchill).
- In continental Europe the perch family includes more than a dozen species including zander and ruffe.

Perch and Man

Although some anglers specifically aim to catch large perch, and specimens up to 2.5kg have been caught in Britain, perch are considered by many fishermen as a by-catch. During the Second World War when food rations were at a premium, small perch from Windermere were canned for human consumption and marketed as 'Perchines'. Although perch are still regarded as good eating, anglers today tend to return them alive to the water.

Further reading: Le Cren, 2001.

Author: Matt Carter, Environment Agency

Zander *Sander lucioperca*

Description

- Resembles a cross between pike and perch, as its alternative name 'Pikeperch' suggests, with two dorsal fins, the first with spiny rays.
- Greenish brown with darkish, broken, vertical bars on flank.
- Large jaws with obvious fangs and numerous small teeth.

Size

Adults usually average 46–66cm in length but can grow to 130cm; average weight around 2.7kg but can weigh up to 15kg.

Biology and behaviour

Spawning takes place from April to June amongst vegetation. The pale yellow eggs are adhesive, sticking to plants and stones, and are guarded by both parents until they hatch five to ten days later. Zander prey mainly on other fishes, normally patrolling open water as solitary individuals, but occasionally in shoals. They actively hunt and pursue prey, unlike pike, which tend to lie in ambush. Their eyes, having a reflective retina, are specially adapted for seeing in water condition where light penetration and visibility are poor.

In their natural range in central and eastern Europe, zander have a much wider choice of prey fish species available than in British waters. This explains why, in Britain, the welfare of stocks of local species can be put at risk by the introduction of zander. Populations of zander can increase quickly, but they have a limited home range. Although the species can become established rapidly in a suitable habitat, further colonization of new reaches of a watercourse is slow unless aided by stocking. The impact on recipient fisheries depends on the rate of expansion of the zander population and the ability of resident prey communities to withstand an additional predator. The fact that zander take relatively small fishes throughout their life-span means that under some circumstances the species can almost annihilate the juvenile component of

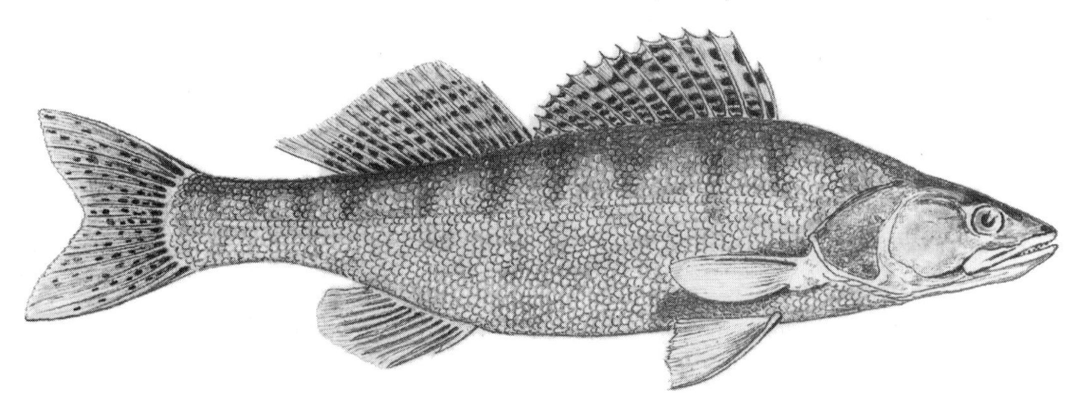

Zander

the prey population. In parts of East Anglia, zander populations have settled into an uneasy balance with native stocks whereas, in the canal systems of the Midlands, this species remains a significant threat to roach populations.

Habitat

Zander inhabit deep, calm waters of lakes, canals and rivers at depths of up to 30m. Their habitat varies according to season, living over pebbles where possible until the winter, when they retreat into pits and trenches to overwinter. In most of their preferred habitats the water has a tendency to be turbid during parts of the year.

Distribution in Britain

The earliest recorded successful introduction of European zander into a British water was in 1878 when 23 individuals were put into two closed lakes on the Woburn Estate in Bedfordshire. The first release into a 'flowing' watercourse took place in 1963, when 97 zander from the Woburn population were used to stock the Great Ouse Relief Channel in Norfolk. After this introduction, zander rapidly colonized the fenland river systems (Great Ouse, Nene and Welland). In 1976 it became evident that zander had been introduced into waters within the catchments of the Rivers Severn and Trent, notably Coombe Abbey Lake, Coventry, and the canal system of the Midlands. More recently this species has been found in south-east England, occurring in the River Thames west of London at Teddington, Molesey and possibly Sunbury, and in three lakes. The increase in the range of zander in Britain since 1972 is clearly demonstrated by the map, despite legislation since 1981, under the Wildlife and Countryside Act and subsequently the Import of Live Fish Act, to control its spread.

World distribution

The original distribution of zander was limited to central and eastern Europe, extending into Russia to the basin of the River Volga. The range of the species now includes southern Scandinavia and western Europe, in particular the Netherlands. Adverse ecological impacts following introduction have been reported in several countries.

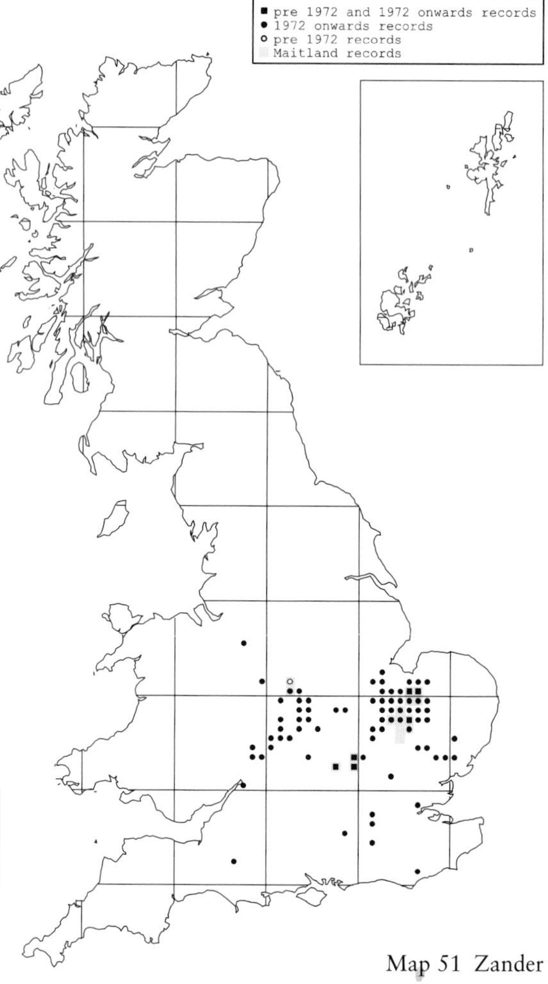

Map 51 Zander

Status

- A non-native species introduced lawfully to a few sites, but then spread by a combination of natural movement and illegal introductions.
- Listed under the Wildlife and Countryside Act 1981 and, in common with other *Stizostedion* species, on the Prohibition of Keeping or Release of Live Fish Orders (see Appendix 3), as a species for which release to the wild is not permitted without a licence.

Hybrids and related species

- Looks similar to the walleye *Stizostedion vitreum* (Mitchill) from North America and early reports sometimes confused the two species.

Zander and Man

The flesh of the zander is whitish and free of bones, tasting delicious whether steamed, baked, fried or cooked in a microwave oven. In England, however, it is surprisingly unpopular as a food fish, whereas in mainland Europe it is an important commercial species being both netted and farmed. Although not universally sought after by anglers, the zander is very highly regarded by devotees for its sporting qualities, with the angling press regularly featuring specimen catches. Unfortunately, some irresponsible enthusiasts are probably guilty of encouraging the illegal spread of zander at the expense of other popular angling species.

Further reading: Lever, 1977.

Author: Phil Hickley, Environment Agency

Grey mullets *Mugilidae*

Description

The following features characterize the three species of grey mullet found in British waters.

- Elongate, sturdy, powerful fish, only slightly laterally flattened.
- Large scales. Blue grey appearance from above, silver on sides, white below.
- Broad, terminal mouth. Teeth small or absent, placed mainly on lips.
- Two dorsal fins: anterior fin with four spines, posterior fin with branched rays like anal fin but shorter. Forked caudal fin.
- Thick-lipped grey mullet *Chelon labrosus* has a protruding, fleshy upper lip and small papillae on the lower edge of the lip. The pectoral fin is relatively long and when folded forward will reach to the eye. The head and eye are relatively small.
- Thin-lipped grey mullet *Liza ramada* has a narrow upper lip with no papillae. The pectoral fin will not reach to the eye when folded forward and there is a dark spot at its base. The head is short, broad and flattened and the eye is moderately large. The pre-orbital bone is serrated.
- Golden grey mullet *Liza auratus* has a similar appearance to the thin-lipped grey mullet. The long pectoral fin will reach to the rear of the eye or beyond when folded forward. The flanks of the fish have a slight golden hue with pronounced golden spots on the cheek and gill cover.

Size

Grey mullet are mature at four or five years and range in size from 40 to 75cm. Thick-lipped grey mullet are the largest of the three species and can grow to specimen weights in excess of 5kg in British waters.

Biology and behaviour

Grey mullet are fast swimming, shoaling fish, which feed by straining organic debris and small organisms from mud. They also browse on algae and take crustaceans and molluscs. They often feed over mud-flats in shallow water at the head of the tide, actively stirring the mud as they do so. They also feed in the

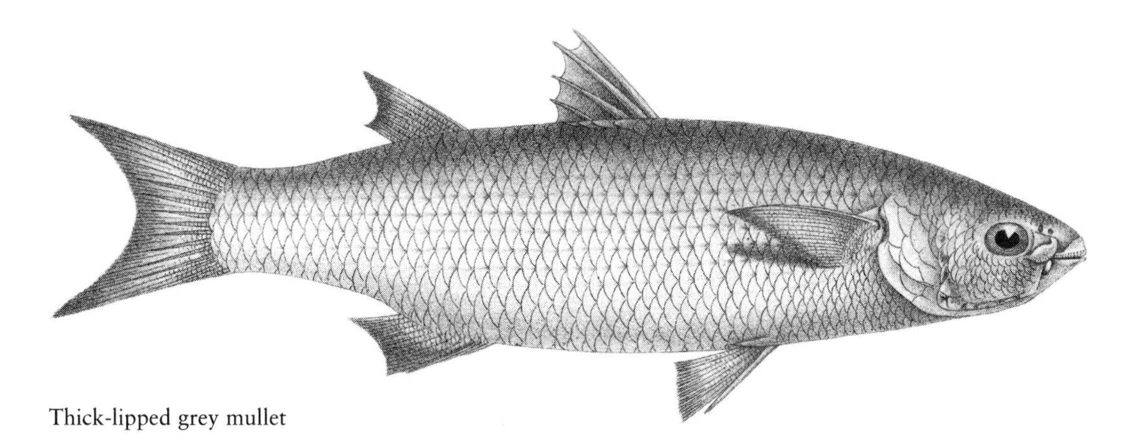

Thick-lipped grey mullet

margins, along walls and piers and at the surface.

Evidence of spawning behaviour is incomplete, but indicates that spawning may take place in coastal waters, on the surface over deep water. The pelagic eggs drift back towards the estuaries from which the adults migrated. Observations on the arrival of very young grey mullet in estuaries and intertidal pools from a range of sites in southern Britain suggest that thick-lipped grey mullet spawn in the spring and thin-lipped in early autumn. In the Thames, very small thin-lipped grey mullet (18–24mm in length) arrive in the estuary as late as October and November and penetrate upstream to Chiswick in West London on the tidal flow.

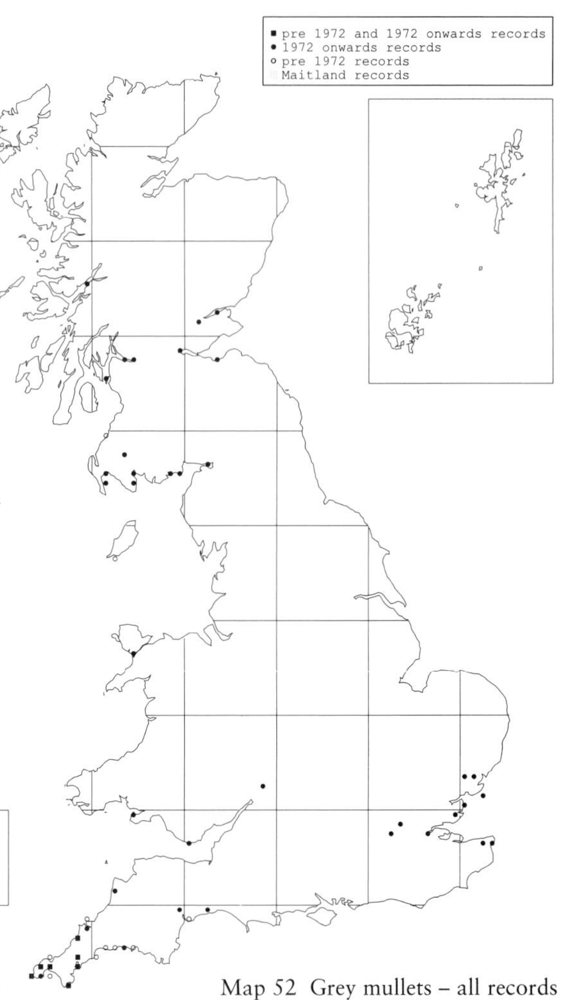

Map 52 Grey mullets – all records

Habitat

Grey mullet migrate northwards into shallow water in the summer months, into estuaries and tidal rivers depending on levels of salinity. They are often found in docks and harbours and near warm water discharges and sewage outfalls. Thick-lipped grey mullet will enter areas with low salinity around the mouth and lower reaches of an estuary. Thin-lipped penetrate far into waters with low salinity, even into freshwater reaches of rivers, often remaining there until the late autumn before returning to deeper coastal waters.

Distribution in Britain

Thick-lipped grey mullet are common inshore and in estuaries around the whole of Britain. Thin-lipped grey mullet have a more southern distribution and are rarely found in northern

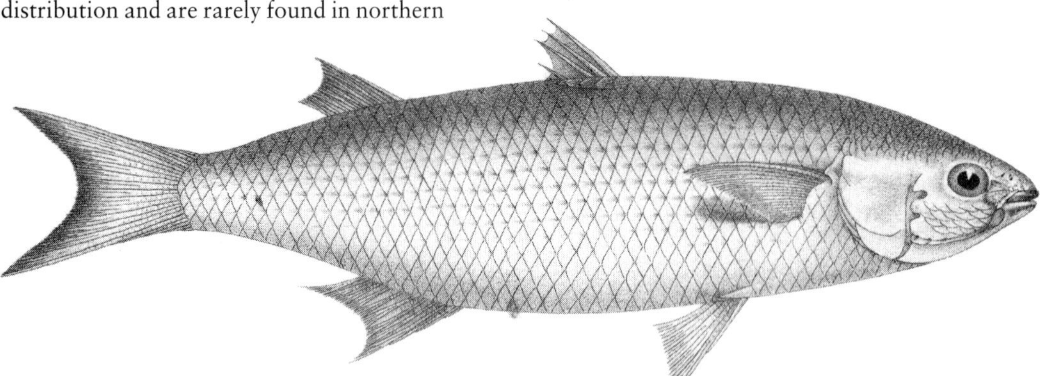

Thin-lipped grey mullet

Scottish waters. Golden grey mullet are the rarest of the three, reported from waters around the whole of Britain, but records may be unreliable due to their close resemblance to thin-lipped grey mullet. Most surveys of freshwater fishes do not attempt to distinguish the individual species of grey mullet and, for this reason, a composite map of all grey mullet records has been included, rather than a separate map for each species.

World distribution

These three species of grey mullet occur in the coastal waters, tidal rivers and estuaries of north-western Europe. Grey mullet are very common fishes in most oceans, and thirteen genera and 70 species have been recognized around the world, occurring mainly in shallow inshore waters in warmer climates. Many species penetrate brackish and fresh waters for short periods and undertake seasonal coastal migrations between estuaries and other transitional waters as both adults and juveniles.

Status

- Not threatened in Britain, the rest of Europe or globally.

Hybrids and related species

- No hybrids are known.
- Although there are no authenticated records, three other species of grey mullet may be encountered around the extreme southern shores of Britain. The sharp-nosed grey mullet *Liza saliens* (Risso), striped or flat-headed grey mullet *Mugil cephalus* Linnaeus and box-lipped grey mullet *Oedalechilus labeo* Cuvier are normally found in warmer waters including the southern Biscay area, around the Iberian Peninsula and in the Mediterranean.

Grey mullet and Man

Because grey mullet are found in inshore areas and estuaries and actively swim in shoals near the surface, humans have exploited them for food, probably for thousands of years. Grey mullet represent important commercial fisheries in the tropics, but less so in northern temperate regions. Fisheries statistics from the Food and Agriculture Organisation of the United Nations indicate that by the early 1960s the average annual world catch of grey mullet from both wild and farmed fisheries was around 60,000 tonnes. In Britain, the commercial interest in grey mullet is small, mainly to serve some ethnic communities for whom grey mullet is a traditional food. Otherwise, habitual British culinary conservatism means that these inexpensive, meaty and very palatable fishes are not commonly offered for sale. In southern areas in the summer months, short seasonal gill- and seine- net fisheries for grey mullet exist, but the total annual catch in Britain is less than 100 tonnes. Grey mullet provide a substantial challenge to anglers in estuaries, harbours and along the coasts, because fine, freshwater tackle is needed to tempt these fastidious, shy and cautious feeders.

Further reading: Colclough *et al.*, 2000; Hickling, 1970; Thomas, 1998; Thomson, 1951, 1953, 1954; Wheeler, 1969, 1979.

Author: Steve Colclough, Environment Agency

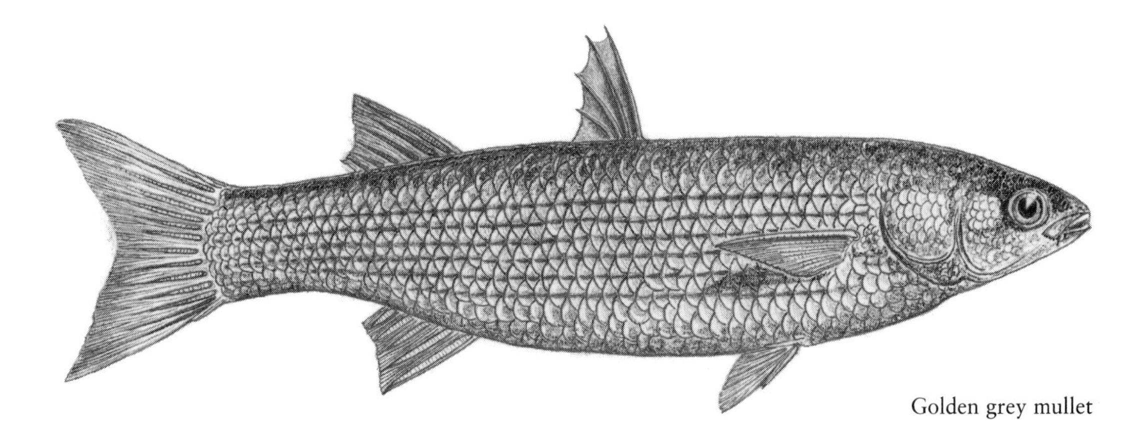

Golden grey mullet

Redbelly tilapia *Tilapia zillii*

Description

- Body deep, with full length dorsal fin, the front half of which is spiny .
- Body olive with six or seven dark vertical bars. Male has red belly in breeding condition.
- Adults show a black, almost circular spot on dorsal fin, the 'tilapia mark', supposedly St Peter's thumb print.

Size

Adults grow to 40cm.

Biology and behaviour

Redbelly tilapia spawn on the sediment of the substrate of a water body and the larvae develop within it. They are usually active only in the daytime, occasionally forming shoals. They feed mainly on water plants, but also take detritus and aquatic invertebrates.

Habitat

In their native range, redbelly tilapia live in tropical fresh and brackish waters. They prefer shallow vegetated areas. Juveniles often occupy seasonal floodplains, which provide excellent conditions for feeding and growth. In Britain, they have become naturalized in the warm outflow from one or two industrial sites.

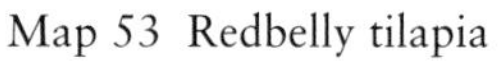
Map 53 Redbelly tilapia

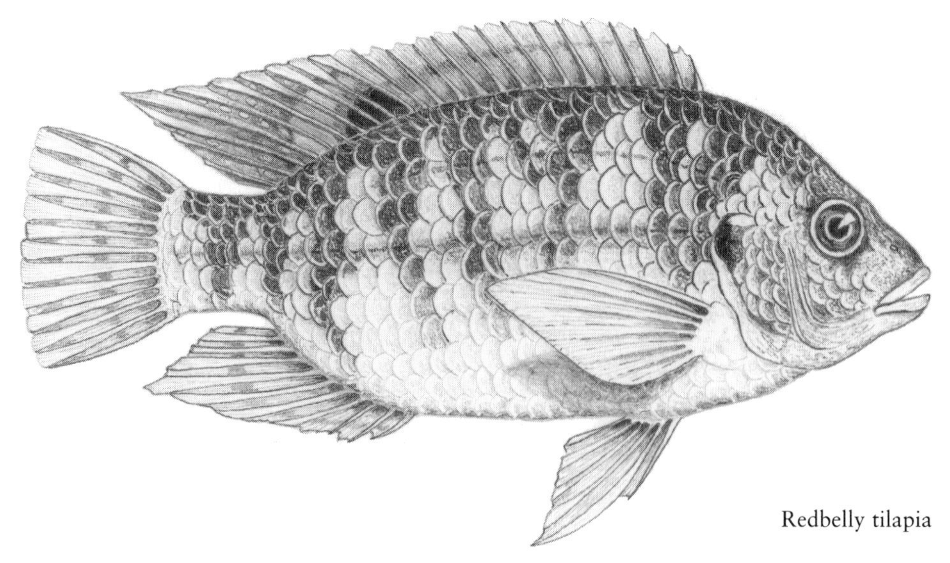
Redbelly tilapia

Distribution in Britain

The first population in the wild in Britain was found in a stretch of canal at Church Street, St. Helens, Lancashire. Specimens were released here in 1963, when a shop selling tropical fish closed down, and they became established as a breeding population in water heated by effluent from a local glassworks. Although water in the canal is still claimed to be warmed by factory effluent, it is now known for its carp fishing rather than for tilapia. Redbelly tilapia have also been recorded from Llyn Trawsfynydd in Wales into which warmed water was discharged from the Trawsfynydd Power Station, which is now decommissioned.

World distribution

The native range of *Tilapia zillii* includes the Niger, Chad and Nile basins of Africa. It has been introduced to other countries such as Sri Lanka and Mexico and many new populations have been established in the wild.

Status

- An introduced, tropical species that needs artificially high water temperatures to survive in Britain and, for this reason, it is unlikely to become invasive here.

Hybrids and related species

- Almost indistinguishable from the redbreast tilapia *Tilapia rendalli* (Boulanger), with which it hybridizes freely.

Redbelly tilapia and Man

Tilapia species in general are exceedingly good to eat and many species are the mainstay of commercial fisheries and fish farms throughout the world although in other areas it is regarded as a nuisance. They feature on ancient Egyptian wall- and tomb-decorations, some dating back as far as the Old Kingdom (over 4000 years ago). The biblical references in the gospels to Jesus feeding the multitude by the Sea of Galilee with a few small loaves and fishes (*Matthew* 14:15–21; 15: 32–39; *Mark* 6:35–44; 8:1–9) would almost certainly have been to *Sarotherodon* (formerly *Tilapia*) *galilaeus* (Linnaeus). Tilapia species are increasingly imported to Britain for sale in supermarkets. Although occasionally kept as an aquarium fish, the redbelly tilapia is not as popular with aquarists as other African cichlids.

Further reading: Lever, 1977.

Author: Phil Hickley, Environment Agency

Flounder *Platichthys flesus*

Description

- Right-eyed flatfish, although left-eyed ('reversed') individuals do occur.
- After flattening, body brownish above with reddish orange blotches and white below.
- Nodules at the base of dorsal and anal rays.
- Single bony ridge between eye and gill cover.
- Scales and general skin surface smooth.
- Anal fin (with 35–46 rays) shorter than dorsal fin (52–64 rays).
- Juveniles superficially similar to juvenile plaice but can be distinguished using fin rays and features on gill cover.

Size

Mature flounder are usually between 20 and 35cm in length and may weigh up to 1.25kg.

Biology and behaviour

Males and females are mature in three and four years respectively. The adults move into coastal waters to spawn, typically at depths of 20 to 40m in the North Sea. The eggs are fertilized in the water column and float around for five to seven days until they hatch. The young are surface-living until they drift inshore and move down to the sea bed when 7–10mm long. When they reach approximately 30mm in length they may move into rivers. Growth is dependent on the population density and availability of food, and the size of individuals between one and two years old can vary considerably.

Flounder are found in groups and, with the exception of their migrations to sea to spawn, they may have a home range limited to as little as 400 square metres. They prey on small fishes and invertebrates, ranging from small crustaceans, such as shrimps and sandhoppers, to worms and bivalve molluscs living in or on the sediment. Small gobies (Gobiidae), another common resident in estuaries, sandeels (Ammodytidae) and bivalves are the principal prey of adults, whilst the younger stages rely mainly on small crustaceans. Their flattened bodies and mottled appearance provide excellent camouflage as they bury themselves in the sand when danger appears.

Habitat

Flounder are bottom-dwelling and are found in a range of river habitats (mainly estuaries and brackish water), and offshore as deep as 50–55m. Spawning may take place in deeper water, particularly in the Baltic Sea, where they spawn at depths as great as 100m. Outside the spawning season, they prefer muddy or sandy estuaries and bays. These areas also provide some protection for the smaller individuals and function as nursery grounds. Flounder move with an incoming tide over newly inundated mud and sand, to take advantage of the feeding areas made available by the tide.

Flounder are capable of surviving in moderately polluted estuaries as well as clean rivers and open coasts. The main limiting factor seems to be the availability of food. They are able to survive water with low oxygen levels, which is a characteristic of many flatfish species, with their large surface to volume ratio. Individuals can remain alive for some hours out of water provided they remain damp. Although they can tolerate very low temperatures, in the winter they will move away from cold fresh water and the upper reaches of estuaries towards the sea.

Distribution in Britain

Flounder are found throughout the coastal waters of Britain. Other than the truly anadromous or catadromous species (for example Atlantic salmon and European eel respectively), the flounder is one of the few fishes, and the only European flatfish, able to live in both fresh and saline water. Although they are frequently found in fresh water, flounder do not have to live in this environment and many do not even migrate upriver, preferring to remain within the tidal reaches of rivers.

The absence of flounder in upland areas results from their need to remain within reach of the coastal waters where the adults breed. The inland limit is approximately 100km from the sea, but flounder are more typically found in the lower freshwater reaches, around 10km from the inland limit of sea water. Populations are unlikely to be common in freshwater reaches where their migratory movement to the sea to spawn is severely impeded. Flounder in Glenfarg

Flounder

Reservoir, Fifeshire, are thought to have been pumped into it with fresh water from the nearby River Earn.

Flounder within estuaries show a marked size distribution, with smaller individuals being located nearer the upper, fresher reaches and larger individuals towards those lower down and more saline. This distribution changes in autumn and winter as a result of the natural migration downstream to warmer water, often assisted by strong river flows.

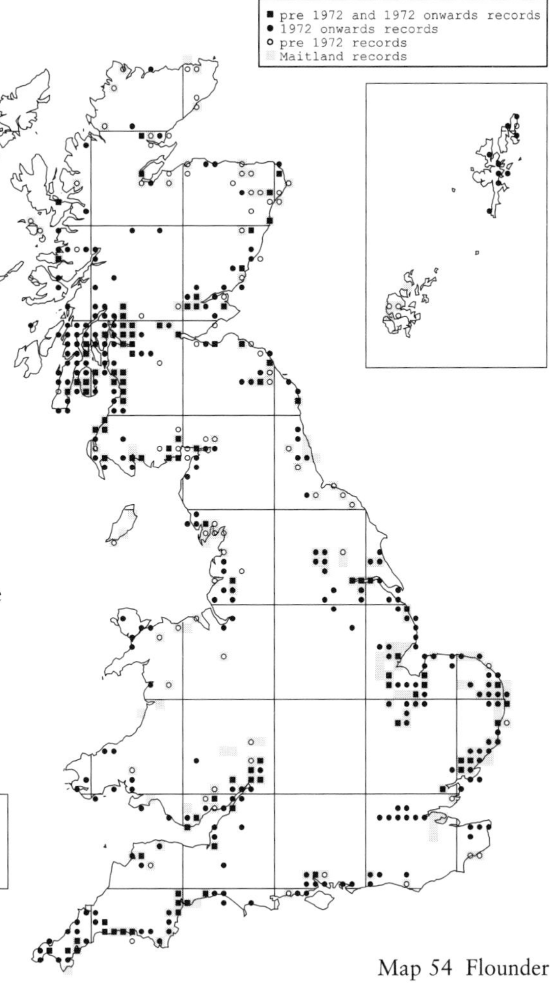

Map 54 Flounder

World distribution

Flounder are found around the coasts of Europe from Norway to the Black Sea, including along the northern Mediterranean coast. They are very common in the Irish, North and Baltic Seas. Populations in warmer climates are less likely than northern populations to move into fresh waters.

Status

- Not threatened in Britain, elsewhere in Europe or globally.

Hybrids and related species

- Hybrids between plaice *Pleuronectes platessa* Linnaeus and flounder have been reported.
- The range of this species possibly overlaps with that of the only other species of flounder in Europe, the Arctic flounder *P. glacialis* (Pallas) in Beloye More (the White Sea) in northern Russia.

Flounder and Man

Flounder are commonly caught by anglers in estuaries and to a lesser extent in fresh waters. They are quite edible but, unlike many other species of flatfish, have not acquired any status as a popular food in Britain. This may be because, in the opinion of some, the flesh is 'muddy' tasting. On the Continent they are more popular as a food and are commercially fished.

Author: Sarah Peaty, Environment Agency

5 Conservation and management of freshwater fishes

Introduction

The publication of this distribution atlas gives a welcome opportunity to summarize the current state of fish conservation in Britain. This chapter reviews their conservation as of 2002, viewing fishes as an essential part of healthy aquatic ecosystems. Because of the large number of organizations and initiatives concerned with the conservation of British freshwater fishes and their habitats, it is possible to discuss only the key aspects here. The references cited will enable more detailed information to be accessed from books and journals, while the Internet sites listed in Appendix 4 are a good source for keeping in touch with the latest developments. Further investigations and analyses of the data compiled for this atlas should be used to support fish conservation in future, and these data will also provide a good baseline for measuring any changes to come.

Fish conservation in Britain

Although Britain has an impoverished fish fauna in comparison with much of Europe, there are many species of conservation interest, either because they are unusual forms, or because they constitute internationally important populations. For example, Britain contains internationally significant populations of Atlantic salmon, twaite shad, bullhead and river lamprey. Glacial relict species and races are also widespread in the mesotrophic and oligotrophic lakes of northern and western Britain: these include Arctic charr, vendace, powan, schelly and gwyniad. Unusual races of brown trout and spineless morphs of three-spined stickleback are also present in some of these sites. One race of trout in northern Scotland, known as the golden trout, is thought to be closely allied to the form that colonized immediately after the last Ice Age.

The conservation of freshwater fishes has been neglected in Britain in comparison with the other vertebrates – mammals, birds, reptiles and amphibians. This may be because fish are less easily observed than many other groups of wildlife, being elusive aquatic animals, which are seldom seen by casual observation. In many circumstances, unlike birds and some mammals, it can be difficult to obtain reliable measures of fish abundance. Freshwater fish conservation in Britain combines two principal requirements: maintaining biodiversity and recreational exploitation. These needs should be complementary, although under some circumstances their objectives may conflict.

Maintaining fish biodiversity

The conservation of populations of rare and threatened native fish species in their remaining locations has been addressed as part of the process of sustaining native biodiversity, using the same conservation techniques applied to many other species. Those bodies that are principally responsible for biodiversity conservation include the statutory nature conservation agencies, English Nature (EN), the Countryside Council for Wales (CCW), Scottish Natural Heritage (SNH), together with their parent Government Departments, the Joint Nature Conservation Committee (JNCC) and the Environment Agency. Non-governmental organizations such as The Wildlife Trusts and the Royal Society for the Protection of Birds also have an important role. These bodies have concentrated on conserving threatened species and on safeguarding the best examples of freshwater habitats, including those given statutory protection, or managed as national or local nature reserves. This combined species and habitats approach is exemplified by the designation of Sites of Special Scientific Interest (SSSIs) (Plates 15 and 16) and Special Areas for Conservation (SACs), as well as by the response of the UK to the Convention on Biological Diversity via the activities under the UK Biodiversity Action Plan (UKBAP).

The 1990s saw two major initiatives that had far-reaching consequences for fish conservation.

The **Habitats and Species Directive** placed an obligation on member states to conserve a wide range of European freshwater fish species (many of which had been previously listed on appendices of the **Bern Convention**). Eight British freshwater fishes listed on annexes of the Directive are both extant and native to the UK. This has resulted in an extensive designation programme of many SACs for these species, which is still in progress. In the case of some anadromous species, such as allis and twaite shad, protection is now afforded from the spawning site to inshore waters, and this legislation has already been successful in preventing construction of a barrage that would have resulted in the destruction of the twaite shad population on the River Usk. All the species covered by legislation are listed in Appendix 2.

The **UK Biodiversity Action Plan** provides a wider and more devolved approach to the conservation of fish species and their habitats. The first four Species Action Plans prepared and published under the auspices of UKBAP were for allis shad, twaite shad, vendace and pollan (the last applying only to Northern Ireland). These four species had been assigned to the Short List of 116 priority species for conservation action on the basis of their international distribution and declining status in the UK. Monitoring populations, researching ecological requirements and protecting habitats that are essential for maintaining populations are actions required for all these species. Reintroduction programmes for vendace are in place in Scotland, where fish from both Cumbrian populations have been translocated to two new locations in the Scottish borders. Two species thought to be extinct in Britain, burbot and houting, were the subject of an Action Plan and a Species Statement respectively within Tranche 2 of the UK Biodiversity Group Action Plans. The burbot may be the subject of a future attempt to reintroduce it to locations within its former range in eastern and southern England. Monitoring rivers where the houting formerly occurred is the only action proposed for this species because it is hoped that natural recolonization may follow after its restoration to key sites in Denmark and Germany. Other species originally included on the UKBAP Middle and Long Lists are now placed on the list of Species of Conservation Concern (SoCC), see Appendix 2.

Recreational exploitation

Native and introduced fish species are exploited by anglers for recreational purposes, supported by the Environment Agency in England and Wales. The Agency has published a good-practice guide to freshwater fisheries and conservation, which brings together advice on both angling and commercial fishing. The principles of managing fisheries in harmony with wildlife conservation are described, as well as current legal requirements and practical techniques for managing flowing and still water habitats. The Environment Agency monitors fisheries, other freshwater organisms and water quality and is responsible for implementing the EU Water Framework Directive.

In Scotland, recreational fisheries (principally Atlantic salmon and sea trout) are managed by District Salmon Fishery Boards. There are 67 Salmon Fishery Districts in Scotland and, for 48 of these, District Salmon Fishery Boards are currently in place. Additional monitoring of salmon, trout and coarse fish fisheries is carried out by the Fisheries Research Services (an agency of the Scottish Executive), Fisheries Trusts and local angling associations. The Scottish Fisheries Co-ordination Centre provides a standard framework for collecting, archiving and accessing information on salmon and other freshwater fish, and on their habitats and fisheries. Water quality is monitored by the Scottish Environment Protection Agency, and the implementation of the EU Water Framework Directive (see p. 146) may mean that this organization will become more actively involved with fisheries management issues in future years.

The techniques for managing fish stocks in Britain include habitat management, fishery regulation and re-stocking of desired species. These approaches to managing fish stocks are intended to maximize the biomass of desired species for exploitation, rather than conserving characteristic assemblages of native fish species, and hence may conflict with the needs of conservation. Even so, there is little doubt that pressure from fishery interests has helped to safeguard many fish populations that might otherwise have been destroyed by pollution or loss of habitat.

These two themes, maintaining fish biodiversity and the recreational exploitation of fish, depend upon freshwater bodies being unpolluted and retaining their full range of structural features and their associated plants and animals. There is thus a common interest in maintaining healthy aquatic environments, which is shared by all those concerned with conserving fish in still or flowing waters.

Fish conservation objectives

A successful strategy for sustaining fish species, as part of Britain's characteristic native biodiversity, should include three key elements.

1 The protection and wise use of aquatic habitats and their associated plant and animal species will conserve the majority of freshwater fishes. This is the preferred means of sustaining fishes and other wildlife in most circumstances, although additional measures may be needed to ensure that the most vulnerable communities survive intact. It is noticeable that the most threatened fishes in Britain tend to be those that have narrowly defined habitat requirements (for example vendace, Arctic charr) or that depend on several different habitats to complete their life history (for example allis and twaite shad). The extinct burbot evidently had multiple habitat requirements, both for living and spawning in rivers, with the young stages preying on invertebrates and the adults on other fish species.

2 The conservation of freshwater fishes in their own right (as individual species, with their intrinsic genetic variation, and as part of natural or near-natural species assemblages) is now rightly getting more attention. This requires a combination of the legislative mechanisms, referred to previously, together with the participation of national and local groups interested in the conservation of freshwater fishes (including those established under the auspices of UKBAP). In most cases, fishes will be conserved alongside other aquatic organisms by general measures designed to maintain unpolluted water and to retain essential habitat features that are needed by fishes. The latter include spawning sites in lakes and rivers, natural river profiles that favour invertebrates for food and aquatic plants that may be eaten by fishes or act as cover against predators. Surveillance, protection and management of important fish populations are necessary to ensure their continued health and survival. In some cases, additional special measures, including captive rearing and release, may be needed to achieve the recovery of populations of extremely rare and threatened species (such as vendace). The conservation of exceptional fish sites, with populations of very localized and threatened species, requires great care and attention because even short periods of pollution, or other adverse factors, can result in the local extinction of such species. Lake species such as Arctic charr, vendace and whitefish are especially vulnerable to this because they are unable to recolonize sites after a local extinction.

3 Some fish species, in common with many other aquatic organisms, are sensitive indicators of water quality and the health of aquatic systems. Thus, surveillance monitoring of fish populations can form an important part of measuring and reporting upon the quality and standards of care applied to lakes and rivers.

In his Environment Agency publication, *Freshwater fisheries and wildlife conservation*, Nick Giles summarizes management issues for freshwater fisheries in relation to wildlife conservation, and his guide includes practical advice for anglers and those responsible for managing flowing and still waters. That publication, and the references it contains, provide examples of good practice for conserving fish in relation to angling and commercial fishing.

Interactions of fishes with other wildlife

Fishes have important interactions with other wildlife. They form part of extensive food chains, both as consumers of other fishes and other vertebrates, invertebrates and plants, and as prey for birds, such as cormorants *Phalacrocorax carbo*, herons *Ardea cinerea* and two sawbilled ducks *Mergus merganser* and *M. serrator*, and for mammals, notably otter *Lutra lutra*. They are also hosts to a variety of diseases and parasites which depend upon the presence of sufficient numbers of their fish hosts for their survival.

The role of fishes in aquatic ecosystems

It is not widely appreciated that fishes, from their position high in the food chain, often act as a 'keystone species' in an ecosystem, causing changes that can affect entire systems, down to the lowest trophic levels. For example, in shallow eutrophic lakes, the introduction of certain fish species often causes a lake to switch from being dominated by pondweeds and other higher plants to domination by phytoplankton. This may occur because bottom-feeding fishes such as bream or carp disturb sediments, releasing nutrients into the water column and physically uprooting plants. Plankton-feeding fishes such as bleak, roach or young bream may have a similar effect, but by a quite different mechanism – by consuming large numbers of cladocerans and other large-bodied zooplankton, they allow the phytoplankton to grow unchecked. No less important is the contribution of migratory fishes such as Atlantic salmon and sea trout to nutrient-poor upland streams. Many individuals die after spawning, and their carcasses are an important source of nutrients for invertebrates and other stream life in these inhospitable environments.

Although some fish species can be regarded as being specialist herbivores (for example spined loach) or piscivores (for example pike), most are generalist feeders that forage opportunistically, consuming whatever prey is available. Prey availability is largely dependent upon use of their habitat by fishes and this can be significantly modified by competitive or predatory interactions, with their own or other species. The community structure and the relative abundance of each species within that community affect the abundance of other wildlife that utilize the aquatic environment. Therefore an understanding of fish populations is important to reveal the health and functioning of aquatic systems.

Fishes as hosts to other organisms

Inevitably, fishes act as hosts to internal and external parasites and to pathogenic organisms such as fungi, bacteria and viruses. The parasites of some species, such as three-spined stickleback, have been studied in detail. The microscopic thin-shelled larvae (glochidia) of freshwater mussels (Margaritiferidae and Unionidae) are parasitic and require a vertebrate host, normally a fish, in which the early development takes place on the gills. As some species of freshwater mussels are now uncommon and potentially threatened in Britain, it is as important to protect the mussels as it is to protect the fishes with which they are associated. This is brought into focus with the internationally protected freshwater pearl mussel *Margaritifera margaritifera*, which uses salmonids, especially salmon and trout, as its glochidial host. A similar association is thought to exist between an even larger and rarer freshwater mussel *M. auricularia* and common sturgeon. Although *M. auricularia* is now extinct in Britain, and common sturgeon no longer breeds here, both have been recorded together as Pleistocene fossils from the River Thames basin, implying that both species probably formerly bred here.

Translocation of fishes

The artificial movement of individual fishes at any life stage (here termed translocation) has important consequences for the health of fish populations and other wildlife. The introduction of fishes from abroad and moving them within Britain is subject to legal controls, with the aims of preventing the establishment of harmful fish species or the arrival and dispersal of fish diseases. The principal legislation deals with keeping or releasing certain non-native fishes under the Import of Live Fish Acts (Scotland, 1978; England and Wales, 1980) and as amended by various Statutory Instruments (see Appendix 3), and the release of fishes into inland waters (Salmon and Freshwater Fisheries Act, 1975; Salmon Act, 1986).

Parasites and diseases have been introduced accidentally with fishes imported from abroad and the consequences can be ecologically and commercially devastating. This was demonstrated by the massive declines of Atlantic salmon in several Norwegian rivers after the unintentional introduction of the parasitic fluke *Gyrodactylus salaris* Malmberg.

While some non-native fish species such as grass carp have the ability to modify aquatic habitats significantly, other introduced species such as zander and Danube catfish can have significant impacts on native fish species. Grass carp have been stocked for 'weed control' and the introduction of this species into sensitive sites, such as plant-rich lakes, can be very damaging. Zander, a large predatory fish, has little direct impact on aquatic environments, but can radically alter the fish community structure in a lake by selectively consuming certain species. Unfortunately, once a non-native species has become established, it can be very difficult to remove, leading to irreversible – and usually undesirable – ecological changes.

To prevent problems arising from the release of non-native fishes in future, a precautionary approach is needed that combines enforceable legislation with codes of conduct for those different sectors that import fish. Such codes of conduct should be underpinned by legislation (via the 'duty of care' concept) to ensure compliance, while codes can be modified to fit changing circumstances more rapidly than legislation. The adoption of a precautionary approach to deal with non-native species is also required to comply with international conventions.

The translocation of native species can also be harmful – sometimes more so than non-native species. For example, introduced ruffe in several lakes are a serious risk to threatened whitefish populations, because they consume the developing eggs in winter.

Statutory conservation responsibilities and legislation in Britain

The Department for Environment, Food and Rural Affairs (DEFRA) has a lead role in formulating environmental policy for the UK, with overall responsibility for safeguarding the countryside and environment of England and Wales. The devolved administrations (National Assembly for Wales and Scottish Executive) have their own responsibilities for sustaining the environment of those countries, including still and flowing freshwater habitats. These Departments deliver their functions via their own staff and through funding statutory agencies in each of the three countries, as described next.

The principal agencies responsible for monitoring and regulating the aquatic environment are the Environment Agency (England and Wales) and the Scottish Environment Protection Agency (Scotland), established by the Environment Act, 1995. These responsibilities include both seeking to prevent pollution and taking action to safeguard freshwater habitats and their associated species. The Environment Agency also has a specific duty to promote fisheries, which is a central theme of their recent report *Our nation's fisheries* reviewing fish stocks in relation to angling and conservation in England and Wales. The functions of these agencies are supported by legislation that deals both with maintaining clean waters and with sustaining healthy fish populations.

European legislation is now assuming greater importance; in the past the **Freshwater Fish Directive** and the **Urban Waste Water Directive** provided for the conservation of freshwater fishes to varying degrees. These Directives are now being subsumed under the **Water Framework Directive**, an ambitious piece of legislation that came into force in 2001, which seeks to bring all European surface waters into 'good ecological quality' or to 'good ecological potential' within 20 years. At the time of writing, this Directive seems likely to have substantial implications for conserving populations of freshwater fishes in Britain and the rest of Europe.

The statutory conservation agencies (EN, CCW, SNH and JNCC) are responsible for implementing international and national obligations for biodiversity conservation. National legislation includes the Wildlife and Countryside Act, 1981, Part 1 of which deals with species conservation matters (including legal protection of animals by listing on Schedule 5, with six freshwater fishes included: sturgeon, allis shad, twaite shad, vendace, powan and burbot), while Part 2 deals with site protection, including the legal basis for designating SSSIs. Amendments to the Wildlife and Countryside Act, 1981, were contained in the Countryside and Rights of Way Act, 2001, for England and Wales, mostly changing clauses of Part 2 of the earlier Act, but also giving a legal status to priority habitats and species as defined in the UK Biodiversity Action Plan. The Habitats and Species Directive, which establishes SACs throughout the European Union, is implemented in Britain via The Conservation (Natural Habitats, &c.) Regulations 1994.

National and international obligations for conserving fish species are summarized in Appendix 2.

The UK government recently carried out a review of the legislation relating to freshwater fishes in England and Wales. This made almost 200 recommendations relating to the management and conservation of fish stocks, including such diverse practices as estuarine netting and agri-environment schemes, and consideration of whether migratory fishes should legally be deemed to be 'freshwater fishes' or not. This has already caused the Environment Agency to develop an Eel Management Strategy, and may bring about new primary legislation relating to fisheries.

Future actions required to conserve freshwater fishes

Hitherto, fish conservation has received less attention than efforts to conserve and manage many other wildlife groups. There is a need to increase awareness of the importance of conserving freshwater fishes as part of maintaining healthy aquatic environments. This is most likely to be achieved by bringing together the major environmental regulation, conservation and angling organizations to work in partnership, together with industrial and recreational users of fresh waters. Their common aim should be to promote fishes as attractive and valuable species, worthy of conservation attention in their own right and as an intrinsic part of healthy fresh waters.

Existing major initiatives, such as developing action plans under UKBAP, strengthening the network of protected sites, obtaining increased funding for agri-environment schemes and implementing European Directives, will all bring significant benefits to fish conservation. As of 2002, the beneficial effects of many of these measures have yet to make a substantial difference to fish conservation in Britain. For example, many candidate SACs are only just beginning to have plans developed for their sympathetic management and ecological restoration. The introduction of the Water Framework Directive is likely to be the major impetus for freshwater conservation, and will shift some attention away from species protection at the highest quality sites towards a broader emphasis on sustaining healthy freshwater habitats and ecosystem functions.

A co-ordinated scheme to carry out surveillance of the most significant fish populations would be very beneficial for fish conservation in Britain, particularly if this could be linked to improved recording of freshwater fishes as part of the follow-up activities to the DAFF project and this publication. Regular reporting of the condition of fish stocks, on threatened fish species and on their critical habitats, should form part of the surveillance of the changes taking place in British biodiversity.

A review of threatened freshwater fishes in Britain is in preparation by JNCC as part of the Species Status project. This will assign conservation statuses, using the latest IUCN Red List Criteria and data derived from the DAFF project. This review should increase awareness of the threats faced by freshwater fishes and help promote actions designed to help recovery of the most endangered species.

Further reading:

General – Baillie & Groombridge, 1996; Lelek, 1980, 1987.

Fish conservation – Anon, 1995, 1998; Giles, 1998; Maitland, 1995; Maitland & Campbell, 1992, 1994.

Interactions with other wildlife – Bean & Winfield, 1995; Carvalho & Moss, 1995; Hastie & Cosgrove, 2001; Moss *et al.*, 1996.

Translocation – Adams & Maitland, 1998; Bullock *et al.*, 1996; Lammens *et al.*, 1992.

Statutory responsibilities – Salmon & Freshwater Fisheries Review Group, 2000.

Appendix 1 Annotated list of publications on the identification and distribution of freshwater fishes in Britain (1686–2004)

Bibliographic details of all the publications included in Appendix 1 are to be found in the Bibliography (pp. 163–169).

Author and date(s)	Short title of publication	Comments
Willughby, 1686	*De historia piscium libri quatuor*	Willughby was a close friend and patron of John Ray who, after Willughby's death in 1672 at the age of 37, saw the book through to publication. It included 187 engravings of fishes, 79 of which were paid for by Samuel Pepys, to whom the work was dedicated.
Williamson, 1740	*The British angler*	Styled as 'a pocket companion for gentlemen fishers', it is poorly illustrated, but is a remarkably comprehensive treatise on many aspects of fishes, fishing and fish husbandry.
Donovan, 1802–08	*The natural history of British fishes*	In five volumes it includes marine fishes, with hand-coloured plates and up to six pages of text on each species. Some species accounts include localities.
Yarrell, 1835–36 (Edn 2, 1841)	*A history of British fishes*	In two volumes, originally published as 19 monthly parts, it includes marine fishes. Illustrated with over 400 woodcuts. The species accounts include variants of English names and some localities.
Hamilton, 1843	*The natural history of British fishes*	In two volumes with 68 colour plates. Later issued as part of Jardine's Naturalist's Library in two editions with conflicting volume numbers. Covers most freshwater species.
Couch, [1860]–65	*A history of the fishes of the British Isles*	In four volumes with 252 coloured plates, it includes marine species. The species accounts include variants of English names and some localities.
Pennell, 1863 (Edn 2, 1875)	*The angler-naturalist: a popular history of British fresh-water fish*	Similar in content and style to Yarrell, which it updates. Includes illustrations from Yarrell.
Buckland, 1873	*Familiar history of British fishes*	Illustrated in black and white, it includes marine species. Republished in 1881 as *Natural history of British fishes*.
Houghton, 1879	*British fresh-water fishes*	Published in two volumes, it includes text for each species and 41 colour plates. Several of the salmonids described are no longer regarded as distinct species. Republished in 1981 in one volume, in a facsimile edition with additional material.
Day, 1880–84	*The fishes of Great Britain and Ireland*	Published in two volumes and illustrated with lithographs of each species (many of which are used in this *Atlas*), it includes marine fishes. Contains much information on the distribution of species, including localities.
Maxwell, 1904	*British fresh-water fishes*	Published in the Woburn Library of Natural History series. Illustrations cover some 20 species.
Regan, 1911	*The freshwater fishes of the British Isles*	Includes 23 'species' regarded by Regan as being 'peculiar to the British Isles' which are merely forms or varieties of accepted species. Includes plates of most species.

Author and date(s)	Short title of publication	Comments
Jenkins, 1925 (Edn 2, 1936)	*The fishes of the British Isles; both fresh water and salt*	Published in the Wayside and Woodland series. Includes marine species, with nearly 300 illustrations of which 128 colour are in colour.
Norman, 1931 (several editions and revisions – see Greenwood, 1963)	*A history of fishes*	General guide to fish biology
Wells, 1941 (several editions and reprints)	*The Observer's Book of freshwater fishes*	Descriptions of 82 species with 76 illustrations, the majority in colour.
Hodgson, 1945 (several editions)	*Freshwater fishes of the British Isles*	A beginners guide with small colour illustrations and short species accounts.
MacMahon, 1946	*Fishlore: British freshwater fishes*	Comprehensive account of fishes and their biology, uses and conservation. Anecdotal accounts of native species, including several segregates of Arctic charr.
Schindler, 1957	*Freshwater fishes*	Thames and Hudson Open Air Guides series translated from a German publication. Includes dichotomous keys to species, brief species accounts and black and white and some colour illustrations of each species.
Greenwood, 1963 (several editions)	*A history of fishes*	Revision of Norman, 1931.
Varley, 1967	*British freshwater fishes: factors affecting their distribution*	A scientific introduction to freshwater fishes. Not an identification guide.
Wheeler, 1969	*The fishes of the British Isles and northwest Europe*	Includes line drawings and maps of the European range of species.
Muus & Dahlstrøm, 1971 (several subsequent editions)	*Freshwater fish of Britain and Europe*	The first English edition, edited by A.Wheeler, translated from the 1967 Danish publication.
Maitland, 1972	*Key to British freshwater fishes*	Dichotomous keys illustrated with drawings and including 10km square distribution maps of each species.
Park, 1972	*Freshwater fishes – A guide to the identification of those found in British waters*	Covers 59 species with a brief account and crude line drawing of each. Many hybrids listed.
Bagenal, 1973	*Identification of British fishes*	Covers marine and freshwater species with keys and simple line drawings.
Maitland, 1977	*The Hamlyn guide to freshwater fishes of Britain and Europe*	See Maitland, 2000.
Wheeler, 1978	*Spotter's guide to fishes*	An illustrated guide for beginners in the Usborne Pocket Books series.
Wheeler, 1978	*Key to the fishes of northern Europe*	Covers more than 350 species, including marine fishes.
Cacutt, 1979	*British freshwater fishes: the story of their evolution*	Species accounts concentrating on historical (palaeontological) records and record specimens. The introductory chapters describe the fossil record of fishes in general.

Author and date(s)	Short title of publication	Comments
Terofal, 1979	*British and European fishes: freshwater and marine species*	Photographs and half-page species accounts covering characteristics, distribution, habits and diet.
Phillips & Rix, 1985	*A guide to the freshwater fish of Britain, Ireland and Europe*	Colour photographic guide to species and their habitats.
Miller, 1986	*A handguide to the fishes of Britain and Europe*	Covers 162 species of common salt- and freshwater fishes with visual key to major groups and colour illustrations. Illustrated by James Nicholls.
Maitland & Campbell, 1992	*Freshwater fishes*	No 75 in the New Naturalist series (hardback and paperback editions). Detailed species accounts, line drawings, black and white and colour photographs, and small scale maps of distribution in Britain and Ireland.
Čihař, 1991	*A guide to freshwater fish*	English edition of a Czech publication with species account, colour plate and European distribution map for each species.
Wheeler & Newman, 1992	*The pocket guide to freshwater fishes of Britain and Europe*	Covers 140 species, each with a page of text and a colour illustration.
Giles, 1994	*Freshwater fish of the British Isles*	Ecological and biological accounts of species with a section of colour photographs covering most species.
Pecl, 1995	*Fishes of lakes and rivers*	Covers 151 species with colour illustrations; descriptions include activity timing, diet, and spawning.
Miller & Loates, 1997	*Fish of Britain and Europe*	Collins Pocket Guide format with colour illustrations, brief species accounts and European distribution maps. Includes marine species.
Wheeler, 1998	*Field key to the freshwater fishes and lampreys of the British Isles*	An AIDGAP guide published as a separate from *Field Studies*. Dichotomous keys to species illustrated with line drawings.
Greenhalgh, 1999	*Freshwater fish*	The natural history of over 160 species with colour illustrations and some photographs and European distribution maps for most species.
Maitland, 2000	*Guide to freshwater fish of Britain and Europe*	Hamlyn Guide series with colour illustrations, brief species accounts and European distribution maps covering over 250 species.
Greenhalgh, 2001	*The pocket guide to freshwater fish of Britain and Europe*	A pocket-sized version of Greenhalgh (1999), with greatly reduced text and no maps.
Pinder, 2001	*Keys to larval and juvenile stages of coarse fishes from fresh waters in the British Isles*	Keys to five immature life stages of 26 species, illustrated with line drawings and colour photographs.
Maitland, 2004	*Keys to the freshwater fish of Great Britain & Ireland.*	Revised and updated version of Maitland, 1972, without detailed distribution maps, but with additional keys.

Appendix 2 Legislation relating to the conservation of freshwater fishes in Britain

The following conservation legislation applies to freshwater fishes in Britain. In each case, the species for which the provisions apply are listed in Table 4 (pp. 152–154), with the appropriate annex or schedule annotated where these apply. Each column in the table is allocated to a legislative instrument. Nomenclature follows that given in the original legislative documents.

EC Directive on the conservation of natural habitats and of wild fauna and flora (Habitats and Species Directive)

Annexes IIa	– designation of protected areas for animal species listed (* = priority species)	species whose natural range includes Great Britain
Annexes IVa	– special protection for animal species listed	
Annexes Va	– exploitation of listed animal species to be subject to management if necessary	

Convention on the Conservation of European Wildlife and Natural Habitats (Bern Convention)

Appendix III – exploitation of listed animal species to be subject to regulation

Convention on International Trade in Endangered Species (CITES)

Appendix I – trade only in exceptional circumstances for species listed
Appendix II – trade in listed species subject to licensing
Appendix III – trade in listed species subject to limited licensing

Wildlife and Countryside Act, 1981

Schedule 5 – animals (other than birds) which are protected
Schedule 9 – animals for which release to the wild is prohibited

The Conservation (Natural Habitats, etc) Regulations, 1994

Schedule 2 – European protected species of animals (natural range includes Great Britain)
Schedule 3 – animals which may not be taken or killed in certain ways

IUCN (1996) Red List of Threatened Animals

Threatened species only cited
CR = critically endangered

UK Biodiversity Action Plan

BP = UK Biodiversity Action Plan Priority Species (Short List)
SoCC = Species of Conservation Concern (formerly included on UK Biodiversity Action Plan Middle or Long Lists)

Table 4 International and national obligations for conserving British freshwater fish

	EC Directive Annexes	Bern Conv Appendix	CITES Appendix	W&C Act Schedules	Cons Regs Schedules	IUCN 1996	UKBAP
Natural range includes GB							
Alosa alosa (allis shad)	IIa,Va	III	-	5 (killing, injuring, taking	3		BP
Alosa fallax (twaite shad)	IIa, Va	III	-	5 (place of shelter)	3		BP
Barbus barbus (barbel)	Va	-	-	-	3		
Cobitis taenia (spined loach)	IIa	III	-	-	-		Was UKBAP Long List, now SoCC
Coregonus albula (vendace)	Va	III	-	5	3		BP
Coregonus autumnalis (pollan) [Ireland only]	Va						BP
Coregonus lavaretus (whitefish)	Va	III	-	5	3		Was UKBAP Long List, now SoCC
Cottus gobio (bullhead)	IIa	-	-	-	-		Was UKBAP Long List, now SoCC
Lampetra fluviatilis (river lamprey)	IIa, Va	III	-	-	3		Was UKBAP Long List, now SoCC
Lampetra planeri (brook lamprey)	IIa	III	-	-	-		Was UKBAP Long List, now SoCC
Petromyzon marinus (sea lamprey)	IIa	III	-	-	-		Was UKBAP Long List, now SoCC

	EC Directive Annexes	Bern Conv Appendix	CITES Appendix	W&C Act Schedules	Cons Regs Schedules	IUCN 1996	UKBAP
Natural range includes GB							
Salmo salar (Atlantic salmon)	IIa, Va in fresh water only	III	-	-	3		Was UKBAP Long List, now SoCC
Salvelinus alpinus (Arctic charr)							Was UKBAP Long List, now SoCC
Thymallus thymallus (grayling)	Va	III	-	-	3		Was UKBAP Long List, now SoCC
Vagrant							
Acipenser sturio (common sturgeon)	*IIa, IVa	III	I	5	2	CR	Was UKBAP Middle List now SoCC
Believed extinct							
Coregonus oxyrinchus (houting) anadromous populations only	*IIa, IVa	III	-	-	-		UKBAP published Species Statement
Lota lota (burbot)							UKBAP published Species Action Plan
Established non-native species listed on Schedule 9 of the Wildlife and Countryside Act, 1981, the release of which into the wild is prohibited without a licence.							
Ambloplites rupestris (rock bass)				9			
Lepomis gibbosus (pumpkinseed, sun-fish or pond-perch)				9			
Micropterus salmoides (large-mouthed black bass)				9			

	EC Directive Annexes	Bern Conv Appendix	CITES Appendix	W&C Act Schedules	Cons Regs Schedules	IUCN 1996	UKBAP
Established non-native species listed on Schedule 9 of the Wildlife and Countryside Act, 1981, the release of which into the wild is prohibited without a licence.							
Rhodeus sericeus (bitterling)	IIa	III		9		-	
Siluris glanis (wels or European catfish)	-	III		9		-	
Stizostedion lucioperca (zander)				9			

Table 5 Special Areas for Conservation (SACs) for fish species in England, Scotland and Wales

* Indicates that some sites are marine sites. Note that some sites are shared between two countries. Many SACs are also notified for more than one species.

	England	Scotland	Wales
Sea lamprey *Petromyzon marinus*	12*	4*	10*
River lamprey *Lampetra fluviatilis*	12*	4*	10*
Brook lamprey *Lampetra planeri*	11	3	6
Allis shad *Alosa alosa*	3*	-	6*
Twaite shad *Alosa fallax*	2*	-	6*
Atlantic salmon *Salmo salar*	10	16	6
Bullhead *Cottus gobio*	14	-	6
Spined loach *Cobitis taenia*	5	-	-

Appendix 3 Legislation relating to the control of non-native species of fish in Britain

The introduction of non-native species of fish (or shellfish) into Britain may have adverse effects upon resident populations of fishes or other organisms, either directly, or indirectly through the impact of the introduction on the aquatic ecosystem. These effects can include:

- direct predation on one or more life stages;
- competition with indigenous fish species for food, cover or spawning sites;
- introduction of new diseases or parasites against which the resident populations have inadequate defences;
- hybridization with resident fishes causing possible reduced viability and fecundity of stocks;
- alteration or degradation of the environment.

Three Acts of primary legislation form the basis for recent measures to licence anyone planning to keep or release any non-native species of fish or shellfish in Britain, whether they are suppliers, dealers, fish farmers or fishery managers.

Import of Live Fish (Scotland) Act 1978 (ILFA Scotland)

This Act gives Ministers power to make an Order to restrict the import, keeping or release of live fish, eggs and gametes of non-native species in Scotland.

Import of Live Fish (England and Wales) Act 1980 (ILFA England and Wales)

This Act gives Ministers power to make an Order to restrict the import, keeping or release of live fish, eggs and gametes of non-native species in England and Wales.

The Wildlife and Countryside Act 1981

Section 14 of this Act makes it an offence to release or allow to escape into the wild any animal which is not ordinarily resident in or a regular visitor to the UK, or which is established in the wild and listed in Schedule 9 of the Act (Table 6A, p. 156), without a licence.

It is impossible to predict precisely how non-native species will perform under new conditions. A precautionary approach has been taken in considering which species may pose a threat if introduced or spread in Britain and which therefore will require a licence. Extensive consultation took place before the following Orders, listed here chronologically, were made under the respective Import of Live Fish Acts. They prohibit the keeping or release of live fish, eggs or gametes of named species, without a licence. The overall aim of these Orders is to afford better protection for native fishes and other species.

Import of Live Fish (Coho salmon) (Prohibition) (Scotland) Order 1980

Covered coho salmon (*Oncorhynchus kisutch*) in Scotland.

Prohibition of Keeping or Release of Live Fish (Pikeperch) (Scotland) Order 1993

Covered zander (pikeperch) in Scotland.

The Prohibition of Keeping or Release of Live Fish (Specified Species) Order 1998

Covered a range of species in England and Wales only.

The Prohibition of Keeping or Release of Live Fish (Specified Species) (Amendment) (England) Order 2003. **Statutory Instrument 2003 No.25.**

Amends the 1998 Order and covers the species listed in Table 6B (pp. 156–158).

The Prohibition of Keeping or Release of Live Fish (Specified Species) (Amendment) (Wales) Order 2003. **Statutory Instrument 2003 No. 416 (W.60)** [quoted verbatim].

Amends the 1998 Order and covers the species listed in Table 6B (pp. 156–158).

The Prohibition of Keeping or Release of Live Fish (Specified Species) (Scotland) Order 2003. Scottish **Statutory Instrument 2003 No. 560.**

This is similar to the amended Order for England and for Wales and includes revocations of earlier Orders in Scotland relating to coho salmon and zander (see above). The most noticeable difference in the species listed for Scotland is the inclusion of ruffe. This species, although apparently native in southern and eastern England, has been introduced elsewhere in Britain either intentionally or accidentally, with particularly damaging consequences in Scotland (see pp. 106, 158).

Further information and advice

Further information about legal responsibilities and constraints on the keeping or release of non-native fish in Britain, including all aspects of licensing can be obtained from the following sources:

Leaflet:

Controls on the keeping or release of non-native fish in England and Wales, published jointly by CEFAS (Centre for Environment, Fisheries and Aquaculture Science) and the Environment Agency;

Website pages:

Controls on the keeping or release of non-native fish in England and Wales at: http://www.efishbusiness.co.uk/controls/part01.asp.

The Prohibition of Keeping or Release of Live fish (Specified Species) (Scotland) Order 2003 Guidance Notes at: http://www.scotland.gov.uk/library5/environment/pkrlf.pdf

Table 6A Species of freshwater fish listed on Schedule 9 under Section 14 the Wildlife and Countryside Act 1981, which it is an offence to release or allow to escape into the wild without a licence. (N.B. The names listed below are as in the published legislation, which does not include scientific names.)

Bass, large mouthed black
Bitterling
Pumpkinseed (Sun Fish or Pond Perch)
Wels, European Catfish
Zander

Table 6B Species of freshwater fish whose keeping or release in any part of England, Wales or Scotland is prohibited except under authority of a Licence granted by the relevant Minister

Common Name	Scientific Name
Asp	*Aspius aspius*
Barbel species	species of the genus *Barbus* (excluding the native *Barbus barbus*)
Bass species (including striped bass, white bass and their crosses e.g. hybrid striped bass)	species of the genus *Morone*
Big-head carp	*Aristichthys /Hypophthalmichthys nobilis*
Bitterling	*Rhodeus sericeus/Rhodeus amarus*
Blacknose Dace	*Rhinichthys atratulus*
Blageon	*Leuciscus souffia*
Blue bream	*Abramis ballerus*
Blue Sucker	*Cycleptus elongatus*

Common Name	Scientific Name
Burbot	*Lota lota*
Catfish	species of the genera *Ictalurus, Ameiurus* and *Silurus*
Charr species (including American Brook Trout)	species of the genus *Salvelinus* (excluding the native *Salvelinus alpinus*)
Chinese black or snail-eating carp	*Mylopharyngodon piceus*
Chinese Sucker also known as Zebra Hi Fin or banded shark/sucker	*Myxocyprinus asiaticus*
Common White Sucker	*Catostomus commersoni*
Danubian bleak	*Chalcalburnus chalcoides*
Danubian Salmon and Taimen	species of the genus *Hucho*
Eastern Mudminnow	*Umbra pygmaea*
European Mudminnow	*Umbra krameri*
Fathead minnow or Roseyreds	*Pimephales promelas*
Freshwater minnow also known as Dragon fish or Pale chub	*Zacco platypus*
Grass carp	*Ctenopharyngodon idella*
Landlocked salmon	non-anadromous varieties of the species *Salmo salar*
Large-mouthed black bass	*Micropterus salmoides*
Marbled trout	*Salmo marmoratus*
Nase	*Chondrostoma nasus*
Northern Redbelly Dace (common minnow)	*Phoxinus/Chrosomus eos*
Pacific salmon and trout (excluding [England and Wales only] rainbow trout, but including steelheads)	species of the genus *Oncorhynchus* (excluding [Scotland only] O. *mykiss and* O. *kisutch*)
Paddlefish	species of the genera *Polyodon* and *Psephurus*

Common Name	Scientific Name
Perch species	species of the genus *Perca* (excluding the native *Perca fluviatilis*)
Pike-perch (including zander)	species of the genus *Stizostedion*
Pike species	species of the genus *Esox* (excluding the native *Esox lucius*)
Red shiner	*Cyprinella/Notropis lutrensis*
Rock bass	*Ambloplites rupestris*
Ruffe [N.B. Scotland only]	*Gymnocephalus cernuus*
Schneider	*Alburnoides bipunctatus*
Silver carp	*Hypophthalmichthys molitrix*
Small-mouth bass	*Micropterus dolomieu*
Snakeheads, Northern or Chinese	species of the genus *Channa* (including *Channa argus*)
Southern Redbelly Dace (common minnow)	*Phoxinus/Chrosomus erythrogaster*
Sturgeon or sterlet	species of the genera *Acipenser, Huso, Pseudoscaphirhynchus* and *Scaphirhynchus*
Sunbleak also known as Sundace, Belica or Motherless Minnow	*Leucaspius delineatus*
Sunfish (including Pumpkinseed, basses, crappies and bluegills)	species of the genus *Lepomis*
Topmouth gudgeon	*Pseudorasbora parva*
Toxostome or French nase	*Chondrostoma toxostoma*
Vimba	*Vimba vimba*
Weather fish	*Misgurnus fossilis*
Whitefish species	species of the genus *Coregonus* (excluding the native *Coregonus lavaretus* and *Coregonus albula*)

Appendix 4 Selected websites with information relating to environmental protection, biodiversity information and fish conservation

Centre for Ecology and Hydrology (CEH)
http://www.ceh.ac.uk
Biological Records Centre (BRC), CEH Monks Wood
http://www.brc.ac.uk

Countryside Council for Wales (CCW)
http://www.ccw.gov.uk

eFishBusiness
http://www.efishbusiness.co.uk

English Nature
http://www.english-nature.org.uk

Environment Agency
http://www.environment-agency.gov.uk

FishBase (global information system on fishes)
http://www.fishbase.org/home.htm
http://www.fishbase.org/search.cfm

Fish Health Inspectorate (part of CEFAS (The Centre for Environment, Fisheries and Aquaculture Science))
http://www.cefas.co.uk/fhi/default.htm

Freshwater Biological Association (FBA)
http://www.fba.org.uk

FRS Freshwater Laboratory, Pitlochry
http://www.frs-scotland.gov.uk

FreshwaterLife (database on freshwater plants and animals)
http://www.freshwaterlife.org

Joint Nature Conservation Committee (JNCC)
http://www.jncc.gov.uk

Life in UK Rivers
http://www.riverlife.org.uk

National Biodiversity Network (NBN)
Main website http://www.nbn.org.uk
Gateway http://www.searchnbn.net

Scottish Environment Protection Agency (SEPA)
http://www.sepa.org.uk

Scottish Fisheries Co-ordination Centre (SFCC)
http://www.sfcc.co.uk

Scottish Executive
http://www.scotland.gov.uk

Scottish Natural Heritage (SNH)
http://www.snh.org.uk

Relevant websites relating to Northern Ireland

Centre for Environmental Data and Recording (CEDaR)
http://www.ulstermuseum.org.uk/cedar/

Environment and Heritage Service (Northern Ireland)
http://www.ehsni.gov.uk

Relevant websites relating to Republic of Ireland

Central Fisheries Board
http://www.cfb.ie

National Parks and Wildlife Service
http://www.duchas.ie/en/NaturalHeritage/NPWSStaffOrganisation/

British records websites

British Record (Rod caught) Fish Committee
http://www.nfsa.org.uk/ntcg/brfc/record_list_coarse_fish.htm
http://www.nfsa.org.uk/ntcg/brfc/record_list_game_fish.htm

Glossary

acclimation – gradual and reversible adjustment of the physiology or morphology of an organism as a result of changing environmental conditions.

ammocoetes – juvenile stage (larvae) of lampreys living in fresh water.

anadromous – migrating from salt water to spawn in fresh water (e.g. salmon).

assemblage – group of individuals, usually of more than one species, occurring together, but not necessarily interacting ecologically.

Balkans – peninsula in southern Europe, between the Black Sea and Aegean Sea to the east, the Adriatic Sea and Ionian Sea to the west and the Mediterranean Sea to the south, including Slovenia in the north-west, Greece in the south and Bulgaria and European Turkey in the east.

barbel – slender, fleshy appendage near the mouth of some bottom-feeding species, used in detecting prey. Also referred to as barbule.

benthic – relating to the bed of a lake or river.

biodiversity – biological diversity: the variety of organisms at all levels, from genetic variants of the same species through species, genera, families and higher taxonomic levels; and the variety of ecosystems, communities and habitats.

biomass – any quantitative estimate of the total mass of organisms comprising all or part of a population or any other specified unit.

Bronze Age – archaeological period approximately 4500 to 2700 years before the present time.

buccal – relating to the mouth or cheek.

by-catch – species caught unintentionally when fishing for other species.

Canestrini, organ of – modified second ray of the pectoral fin in male spined loach, thought to be used during courtship and mating.

catadromous – migrating from fresh water to spawn in sea water (e.g. European eels).

chironomids – small flies (Diptera), often referred to as non-biting midges, with aquatic larvae that are a common prey item for fishes.

cichlid – member of the mainly tropical family, Cichlidae.

circumboreal – distributed around the high latitudes of the northern hemisphere.

coarse fishes – usually taken to mean freshwater species not included among '**game fishes**' (q.v.).

cobble – type of substrate or sediment particle (e.g. in a river) between 64 and 256mm in diameter (cf. **gravel**).

coble – an open flat-bottomed fishing boat.

copepods – small crustaceans, often occurring in very large numbers, that are commonly preyed upon by fishes.

coregonid – member of the whitefish family, Coregonidae.

cyprinid – member of the carp family, Cyprinidae.

depauperate – impoverished.

elver – immature European eel at the stage when it arrives at our coasts and first enters fresh water.

eutrophic – water that is rich in the nutrients required by green plants. Particularly in summer, this can lead to rapid plant growth and associated decay leading to depletion of dissolved oxygen levels, especially in still waters; hence eutrophication, the process of nutrient enrichment (cf. **mesotrophic**, **oligotrophic**).

fins – several types of fin are referred to in the text. Fins take many forms and not all species have all types.

- **adipose** – single, small, fleshy fin without rays, on back between dorsal fin and tail.
- **anal** – single fin on belly between vent (anus) and tail.
- **caudal** – tail fin (see also **wrist**).
- **dorsal** – main fin on the back. Some species have a double dorsal fin, one behind the other, both with rays (cf. **adipose fin**), although in some species (e.g. ruffe) they may be fused together.
- **pectoral** – paired fins behind gill covers.
- **pelvic** – paired fins on belly in front of anal fin and vent.

fry – young fishes that have recently hatched from eggs, excluding those classed as larvae.

fyke net – large hoop net that acts as a funnel to trap swimming fishes.

game fishes – usually taken to mean salmon, trout and other **salmonids** (q.v.).

gill net long nets suspended in the water column, normally with floats on the surface and weights at the bottom, designed to catch fishes that swim into the fine mesh and become caught by their gills.

gill rakers – comb-like structures on the gills, found in some species (e.g. shads) that feed on plankton.

glochidia – minute, larval stage of large, aquatic bivalve molluscs. They are parasitic on fishes, normally attaching to the gills.

gravel – type of substrate or sediment particle of mineral origin between 2 and 64mm in diameter, sometimes further divided into 'large' or 'small' gravel but without defined size limits (cf. **cobble**).

gynogenesis – process in which an egg develops parthenogenetically having being activated but not fertilized by sperm.

Holarctic – terrestrial zoogeographical region including most of the northern hemisphere from the Pole to the Tropic of Cancer.

Iberian Peninsula – the landmass comprising modern Spain and Portugal.

indigenous – belonging naturally to an area or country.

invertebrate – any animal without a backbone (spine), including insects, arachnids, crustaceans and molluscs.

Iron Age – archaeological period approximately 2700 to 1950 years before the present time.

larvae – young fishes that have recently hatched from eggs, but which have not yet developed into the typical form of the fish (see also **fry**). Larvae are usually characterized by transparency, absence of scales and embryonic, undifferentiated fins, and a yolk sac may still be present.

littoral zone – shore zone, used here to mean the shore of a lake and including shallow water.

macrophytes – aquatic higher plants (flowering plants and ferns).

Mesolithic – archaeological period approximately 12,000 to 6500 years before the present time.

mesotrophic – water with intermediate levels of the nutrients required by green plants (cf. **eutrophic, oligotrophic**).

morphology – form and structure of an organism, particularly the external features; hence **morphologically** – relating to the morphology.

natal river – river in which an anadromous fish hatched and lived before going to sea.

Neolithic – archaeological period approximately 6500 to 4000 years before the present time.

oligotrophic – water with low levels of the nutrients required by green plants, where dissolved oxygen levels are not depleted in summer (cf. **eutrophic, mesotrophic**).

operculum (pl. **opercula**; adj. **opercular**) – gill cover.

Palaeolithic – archaeological period that ended approximately 10,000 years before the present time.

papillae – small conical projections from the body surface.

parr – young trout or salmon lacking mature coloration, with regular dark blotches (parr marks) along the sides. See also **smolt**.

pathogenic – producing, or capable of producing, disease.

pelagic zone – water column of a still or flowing water body, away from the littoral zone and above the benthic zone; open water.

pharyngeal teeth – tooth-like structures in the gullet or anterior part of the gut.

phylogenetic – the theoretical evolutionary relationships within and between groups, such as between two species.

piscivorous – feeding on fishes.

phytoplankton – plant life (normally algae) that is **planktonic** (q.v.).

plankton (adj. **planktonic**) – organisms that are unable to maintain their position or distribution independent of the movement of water. See **phytoplankton, zooplankton**.

Pleistocene – geological term for the main part of the **Quaternary** period (q.v.) during which successive cold glaciations and warmer interglacial episodes occurred, but excluding the present interglacial, known as the Holocene.

plumb-lining – fishing by trolling (towing) spinners (artificial lures) above a heavy keeled lead weight behind rowing boats, at depths of up to 20m.

profundal zone – deepest zone of a lake below the level of effective light penetration and hence of vegetation.

put-and-take – stocking with fishes bred and reared in captivity to a size suitable to be taken by angling.

Quaternary – geological term for the most recent period (roughly 1.6 million years), which includes the **Pleistocene** (q.v.) and the present interglacial, known as the Holocene.

redd – excavation, normally in a gravel substrate, used by many salmonid fishes to protect incubating eggs.

relict – remnant (e.g. a species) that is presumed or known to have been more widespread formerly, now persisting only in isolated areas or habitats.

riffle – fast water section of a stream or river, with a broken surface, where shallow water flows over stones or gravel.

riparian – relating to a river bank.

rotifers – minute, primitive, unsegmented animals, sometimes referred to as 'wheel animalcules', which often occur in large numbers and are preyed upon by fishes.

salmonid – member of the family Salmonidae, including salmon, trout, charr, grayling and whitefish.

sawbill duck – duck of the genus *Mergus*, including goosander *Mergus merganser* and red-breasted merganser *Mergus serrator*.

Seine net – fishing net for encircling, with floats at the top and weights at the bottom.

smolt – 'silvery' immature salmonid at the stage that it migrates from fresh water to the sea.

spate river – river subject to regular, seasonal fluctuations in water levels and flows with considerable periods of high levels and flows fed primarily by rainfall or snow melt.

spawn – to lay eggs, or deposit milt (semen).

specimen fish – angling term for a large fish approaching the maximum size that might be expected.

steelhead – the anadromous form of rainbow trout *Oncorhynchus mykiss*.

substrate – sediment or other surface at the bottom of the water column. Also referred to as **substratum**.

symbiotic – a relationship in which two or more species live together usually to their mutual benefit.

taxonomic – relating to the classification and naming of species.

trophic status – relating to the productivity of an ecosystem, see **eutrophic, mesotrophic, oligotrophic**.

tundra – biome typified by permanently frozen subsoil, an absence of trees and a vegetation composed mainly of mosses and lichens.

vertebrate – animal with a backbone (spine): fishes, amphibians, reptiles, birds and mammals.

water column – the body of still or flowing water between the surface and the substrate or bottom.

whitefish – one or more of the *Coregonus* species or, if referring to a single species, *Coregonus lavaretus*.

wrist – the narrow junction of the main body and the tail (caudal fin), otherwise known as the caudal peduncle.

zooplankton – very small animals living in large numbers, e.g. copepods and rotifers, which are **planktonic** (q.v.).

Bibliography and further reading

Adams, C. E. & Maitland, P. S., 1998. The Ruffe population of Loch Lomond, Scotland: its introduction, population expansion and interaction with native species. *Journal of Great Lakes Research* **24**: 249–262.

Aldridge, D. C., 1997. *Reproductive ecology of bitterling* (*Rhodeus sericeus* Pallas) *and unionid mussels*. Unpublished Ph.D. thesis, University of Cambridge.

Aldridge, D. C., 1999. Development of European bitterling in the gills of freshwater mussels. *Journal of Fish Biology* **54**: 138–151.

Anon., 1995. *Biodiversity: the UK Steering Group Report*. London: HMSO.

Anon., 1998. *UK Biodiversity Group Tranche 2 Action Plans: volume VI – terrestrial and freshwater species and habitats*. Peterborough: English Nature.

Anon., 2003. *Scottish salmon and trout catches 2002*. Fisheries Research Services Statistical Bulletin, Fisheries Series 2002 catches (Fis/2003/1). Edinburgh: Scottish Executive, Environment & Rural Affairs Department.

Arme, C. & Owen, R.W. 1964. Massive infections of the three-spined stickleback with *Schistocephalus solidus* (Cestoda: Pseudophyllidae) and *Glugea anomala* (Sporozoa: Microsporidia). *Parasitology* **54**: 10–11.

Ayres, K., Locker, A. & Serjeantson, D. 2003. The medieval abbey: food consumption and production. *In* Hardy, A., Dodd, A. & Keevill G. (Eds), *Aelfric's Abbey: Excavations at Eynsham Abbey, Oxfordshire 1989–1992* (Thames Valley Landscapes Monograph 16). Oxford: Oxford Archaeology.

Backiel, T. & Zawiska, J., 1968. Synopsis of the biological data on the bream, *Abramis brama* (L.). *F.A.O. Fisheries Synopsis No. 36*. Rome: UNESCO, F.A.O.

Baglinière, J. L. & Elie, P. (Eds), 2000. *Les aloses* (*Alosa alosa* et *Alosa fallax* spp.) *Écobiologie et variabilité des populations*. Paris: INRA-CEMAGREF.

Baillie, J. & Groombridge, B. (Eds), 1996. *The IUCN Red List of Threatened Animals*. Gland: IUCN.

Balon, E. K. (Ed.), 1980. *Charrs – salmonid fishes of the Genus* Salvelinus. The Hague: Dr W. Junk.

Baroudy, E. 1995. Arctic charr (*Salvinus alpinus*) in Windermere (Cumbria). *Freshwater Forum* **5**: 185–192.

Beamish, F. W. H., 1980. Biology of the North American anadromous sea lamprey, *Petromyzon marinus*. *Canadian Journal of Fisheries and Aquatic Sciences* **37**: 1924–1943.

Bean, C. W. & Winfield, I. J., 1995. The habitat use and activity patterns of roach (*Rutilus rutilus* (L.)), rudd (*Scardinius erythrophthalmus* (L.)), perch (*Perca fluviatilis* L.) and pike (*Esox lucius* L.) in the laboratory: the role of predation threat and structural complexity. *Ecology of Freshwater Fish* **4**: 37–46.

Bergstedt, R. A. & Seelye, J. G., 1995. Evidence for the lack of homing by sea lampreys. *Transactions of the American Fisheries Society* **124**: 235–239.

Berners, Dame Juliana (attributed), 1496. *The treatyse of fysshynge wyth an angle*. In: *The boke of St Albans*. Wynken de Worde. London.

Bianco, P. G. & Muciaccia, M., 1982. Primo reporto di *Lampetra fluviatilis* per l'Adriatico. *Natura (Milano)* **73**: 155–158.

Billard, R., 1997. *Les Poissons d'eau douce des rivières de France*. Lausanne: Delachaux et Niestlé.

Bogutskaya, N. G. & Naseka, A. M., 1996. *Cyclostomata and fishes of Khanka Lake drainage area (Amur river basin). An annotated check-list with comments on taxonomy and zoogeography of the region*. St Petersburg: Zoological Institute of the Russian Academy of Sciences.

Broughton, R., 2000. *The complete book of the grayling*. London: Robert Hale.

Buchan, J., 1927. *John Macnab*. London: Hodder & Stoughton. (Many subsequent editions.)

Buckland, F., 1873. *Familiar history of British fishes*. London: Society for Promoting Christian Knowledge. (Republished, 1881, under the title *The natural history of British fishes*.)

Bullock, J. M., Hodder, K. H., Manchester, S. J. & Stevenson, M. J., 1996. *Review of information, policy and legislation on species translocation*. JNCC Report No. 261. Peterborough: JNCC.

Cacutt, L., 1979. *British freshwater fishes: the story of their evolution*. London: Croom Helm.

Carvalho, L. & Moss B., 1995. The current status of a sample of English Sites of Special Scientific Interest subject to eutrophication. *Aquatic Conservation: Marine and Freshwater Ecosystems* **5**: 191–204.

Čihař, J., 1991. *A guide to freshwater fish*. London: Treasure Press.

Clarke, D. V., 1998. The environment and economy of Skara Brae. *In* Lambert, R. A. (Ed.), *Species history in Scotland*, pp. 8–19. Edinburgh: Scottish Cultural Press.

Coad, B. W., 1981. A bibliography of the sticklebacks (Gasterosteidae: Osteichthyes). *Syllogeus* **35**: 1–142.

Colclough, S. R., Dutton, D., Cousins, T. & Martin, A., 2000. *A fish population survey of the tidal Thames*. Unpublished report to the Environment Agency.

Coles, J. M., 1971. The early settlement of Scotland: excavations at Morton, Fife. *Proceedings of the Prehistory Society* **37**: 284–366.

Coles, J. M., 1987. Animal bones from Meare Village East 1932–1956. *Somerset Levels Papers* **13**: 230–232.

Cooper, J. A. & Chapleau, F., 1998. Monophyly and intrarelationships of the family Pleuronectidae (Pleuronectiformes), with a revised classification. *Fishery Bulletin* **96**(4): 686–726.

Corbet, G. B. & Harris, S. (Eds), 1991. *The handbook of British mammals*. Oxford: Blackwell.

Couch, J., [1860]–65. *A history of the fishes of the British Isles*. London: Groombridge.

Cowx, I. G., 2001. *Factors influencing coarse fish populations in rivers*. Bristol: Environment Agency.

Craig-Bennett, A., 1931. The reproductive cycle of the three-spined stickleback, *Gasterosteus aculeatus* L. *Philosophical Transactions of the Royal Society of London, B* **219**: 197–279.

Crisp, D. T., 2000. *Trout and salmon: ecology, conservation and rehabilitation*. London: Fishing News Books.

Darby, H. C., 1940. *The medieval fenland*. Cambridge: Cambridge University Press.

Dartnall, H. J. G., 1973. Parasites of the nine-spined stickleback *Pungitius pungitius* (L). *Journal of Fish Biology* **5**: 505–509.

Day, F., 1880. *The fishes of Great Britain and Ireland*, 2 vols. London: Williams & Norgate.

Donoghue, S., 1988. *Gasterosteus aculeatus*, a new host for the metacercariae of *Cryptocotyle concavum* (Digenea). *Journal of Fish Biology* **33**: 657–658.

Donovan, E., 1802–08. *The natural history of British fishes*, 5 vols. London: privately published.

Elkan, E., 1962. *Dermocystidium gasterostei* n. sp., a parasite of *Gasterosteus aculeatus* L. and *Gasterosteus pungitius* L. *Nature* **196**: 958–960.

Elliott, J. M., 1994. *Quantitative ecology and the brown trout*. Oxford: Oxford University Press.

Ellison, N. F. & Chubb J. C., 1968. The smelt of Rostherne Mere, Cheshire. *Lancashire and Cheshire Fauna Society, Occasional Publication* **53**: 7–16.

English Heritage, 1999. *Archaeological Review 1997–98*. London: English Heritage.

Farr-Cox, F., Leonard, S. & Wheeler, A., 1996. The status of the recently introduced fish *Leucaspius delineatus* (Cyprinidae) in Great Britain. *Fisheries Management and Ecology* **3**: 193–199.

Farrar, L., 1998. *Ancient Roman gardens*. Stroud: Sutton Publishing.

FishBase: Froese, R. &. Pauly, D. (Eds), 2003. FishBase. World Wide Web electronic publication. www.fishbase.org .

Fitter, R. S. R., 1945. *London's natural history* (The New Naturalist Library No. 3). London: Collins.

Frear, P. A. & Shannon, J. C., 1994. Recent occurrences in Yorkshire of sea lampreys *Petromyzon marinus* (Petromyzontidae). *Naturalist* **119**: 43.

Froese, R. & Pauly, D. (Eds), 1998. *FishBase 1998: concepts, design and data sources*. Manila, Phillipines: ICLARM.

Frost, W. E., 1974. *A survey of the rainbow trout in Britain and Ireland*. London: Salmon & Trout Association.

Frost, W. E. & Brown, M. E., 1967. *The trout* (New Naturalist Monograph No. 21). London: Collins.

Froufe, E., Magyary, I., Lehoczky, I. & Weiss, S., 2002. mtDNA sequence data supports an Asian ancestry and single introduction of the common carp into the Danube basin. *Journal of Fish Biology* **61** (1): 301–304.

Gelling, M., 1984. *Place-names in the landscape*. London: Dent.

Giles, N., 1994. *Freshwater fish of the British Isles*. Shrewsbury: Swan Hill Press.

Giles, N., 1998. *Freshwater fisheries & wildlife conservation: a good practice guide*. Bristol: Environment Agency.

Gozlan R. E., Pinder A. C. & Shelley J., 2002. Occurrence of the Asiatic cyprinid *Pseudorasbora parva* in England. *Journal of Fish Biology* **61**(1): 298–300.

Gozlan, R. E., Pinder, A. C., Durand, S. & Bass, J., 2003. Could the small size of *Leucaspius delineatus* be an ecological advantage in invading British watercourses? *Folia Zoologica* **52** (1): 99–108.

Greenhalgh, M., 1999. *Freshwater fish*. London: Mitchell Beazley.

Greenhalgh, M., 2001. *The pocket guide to freshwater fish of Britain and Europe*. London: Mitchell Beazley.

Greenwood, P. H., 1963. *A history of fishes*. London: Ernest Benn. (Revised edition of Norman, J. R., 1931.)

Gunderson, J. L., Klepinger, M. R., Bronte, C. R. & Marsden, J. E., 1998. Overview of the International Symposium on Eurasian Ruffe (*Gymnocephalus cernuus*). Biology, impacts, and control. *Journal of Great Lakes Research* **24**(2): 165–169.

Hamilton, R., 1843. *The natural history of British fishes* **2**. Edinburgh: W. H. Lizars. (Reissued in 1845–46 as *Ichthyology*, Vol. VI in Jardine's Naturalist's Library.

Hänfling, B., Hellemans, B., Volckaert, F. A. M. & Carvalho, G. R., 2002. Late glacial history of the cold-adapted freshwater fish *Cottus gobio*, revealed by microsatellites. *Molecular Ecology* **11**: 1717–1729.

Hansen, M. M., Mensberg, K.-L. D. & Berg, S., 1999. Postglacial recolonization and genetic relationships among whitefish (*Coregonus* sp.) populations in Denmark, inferred from mitochondrial DNA and microsatellite markers. *Molecular Ecology* **8**: 239–252.

Hardisty, M. W. & Potter, I. C. (Eds), 1971. *The biology of the lampreys* **1**. London: Academic Press.

Hardy, E., 1954. The bitterling in Lancashire. *Salmon & Trout Magazine* **142**: 548–553.

Hastie, L. C. & Cosgrove, P. J., 2001. The decline of migratory salmonid stocks: a new threat to pearl mussels in Scotland. *Freshwater Forum* **15**: 85–96.

Hickling, C. F., 1970. A contribution to the natural history of the English grey mullets [Pisces, Mugilidae]. *Journal of the Marine Biological Association of the UK* **50**: 609–633.

Hodgson, N. B., 1945. (Edn 2, 1949). *Freshwater fishes of the British Isles*. London: Eyre & Spottiswood.

Holčik, J. (Ed.), 1986. *The freshwater fishes of Europe I/II. Petromyzontiformes*. Wiesbaden: Aula-Verlag.

Holčik, J. & Mihalik, S., 1968. *Freshwater fishes*. London: Spring Books.

Houghton, W., 1879. *British freshwater fishes*. London: Mackenzie.

Hubbs, C. L. & Potter, I. C., 1971. Distribution, phylogeny and taxonomy. *In* Hardisty, M. W. & Potter I. C. (Eds), *The biology of lampreys* **1**: 1–65. London: Academic Press.

Huet, M., 1949. Aperçu des relations entre la pente et les populations piscicoles des eaux courantes. *Schweizerische Zeitschrift für Hydrologie* **11**: 333–351.

Huet, M., 1959. Profiles and biology of western European streams as related to fish management. *Transactions of the American Fisheries Society* **88**: 155–163.

Hughes, T., 1983. *River*. London: Faber & Faber.

Hutchinson, P. & Mills, D. H., 1987. Characteristics of spawning-run smelt, *Osmerus eperlanus*, from a Scottish river, with recommendations for their conservation and management. *Aquaculture and Fisheries Management* **18**: 249–258.

Hynes, H. B. N., 1970. *The ecology of running waters*. Liverpool: Liverpool University Press.

Jeffries, M. & Mills, D. [H.], 1990. *Freshwater ecology*. London: Belhaven Press.

Jenkins, J. T., 1925. (Edn 2, 1936). *The fishes of the British Isles, both fresh water and salt*. London: Warne.

Jerome, J. K., 1889. *Three men in a boat*. Bristol: Arrowsmith. (Many subsequent editions.)

Katano, O. & Maekawa, K., 1997. Reproductive regulation in the female Japanese minnow, *Pseudorasbora parva* (Cyprinidae). *Environmental Biology of Fishes* **49**: 197–205.

Keith, P., 1995. *Cinq examples d'évolutions de populations piscicoles*. France: RNDE, SFF-MNHN, CSP, Ministère de l'Environment.

Kennedy, M. & Fitzmaurice, P., 1970. The biology of the tench *Tinca tinca* (L.) in Irish waters. *Proceedings of the Royal Irish Academy* (B) **69**: 31–82.

Kipling, C., 1972. The commercial fisheries of Windermere. *Transactions of the Cumberland and Westmorland Antiquarian and Archaeological Society*, N.S. **72**: 156–204.

Kottelat, M. (Ed.), 1997. European freshwater fishes. *Biologia* **52** (Supplement 5)1–271.

Lamb, H. H., 1995. *Climate, history and the modern world*. London: Routledge. (A fully revised edition of his 1982 publication with same title, published by Methuen.)

Lammens, E. H. R. R., Geursen, J. & MacGillavry, P. J., 1987. Diet shifts, feeding efficiency and coexistence of bream (*Abramis brama*), roach (*Rutilis rutilis*) and white bream (*Blicca bjoerkna*) in hypertrophic lakes. *In* Kullander, S. O. & Fernholm, B. (Eds.), *Proceedings of the Fifth Congress of European Ichthyology* (Stockholm, 1985), pp. 153–162. Stockholm: Swedish Museum of Natural History.

Lammens, E. H. R. R., Frank-Landman, A., McGillavry, P. J. & Vlink, B., 1992. The role of predation and competition in determining the distribution of common bream, roach and white bream in Dutch eutrophic lakes. *Environmental Biology of Fishes* **33**: 195–205.

Langford, T. E., 1981. The movement and distribution of sonic-tagged coarse fish in two British rivers in relation to power station cooling-water outfalls. *In* Lang, F. M. (Ed.), *Proceedings of the Third International Conference on Biotelemetry*, pp. 197–232. Laramie: University of Wyoming.

Le Cren, E. D., 2001. The Windermere perch and pike project: an historical review. *Freshwater Forum* **15**: 3–34.

Lelek, A., 1980. *Threatened freshwater fishes of Europe*. Nature and Environment Series No. 18. Strasbourg: Council of Europe.

Lelek, A., 1987. *The freshwater fishes of Europe* **9**: *Threatened fishes of Europe*. Wiesbaden: Aula-Verlag.

Lever, C., 1977. *The naturalized animals of the British Isles*. London: Hutchinson.

Li, W., Scott, A. P., Siefkes, M. J., Yan, H., Liu, Q., Yun, S.-S. & Cage, D. A., 2002. Bile acid secreted by male sea lamprey that acts as a sex pheromone. *Science* **296**: 138–141.

Lucas, M. C. & Baras, E., 2001. *Migration of freshwater fishes*. London: Blackwell Science.

MacCrimmon, H. R., 1971. World distribution of rainbow trout (*Salmo gairdneri*). *Journal of the Fisheries Research Board of Canada* **28**: 663–704.

MacCrimmon, H. R. & Campbell, J. S., 1969. World distribution of brook trout, *Salvelinus fontinalis*. *Journal of the Fisheries Research Board of Canada* **26**: 1699–1725.

MacMahon, A. F. M., 1946. *Fishlore: British freshwater fishes*. Harmondsworth: Penguin Books.

Maekawa, K., Iguchi, K. & Katano, O., 1996. Reproductive success in male Japanese minnows, *Pseodorasbora parva*: observations under experimental conditions. *Ichthyological Research* **43** (3): 257–266.

Maitland, P. S., 1969. A preliminary account of the mapping of the distribution of freshwater fish in the British Isles. *Journal of Fish Biology* **1**: 45–58.

Maitland, P. S., 1972. A key to the freshwater fishes of the British Isles with notes on their distribution and ecology. *Scientific Publication* No. 27. Windermere: Freshwater Biological Association.

Maitland, P. S., 1977. *Freshwater fish of Britain and Europe* (Hamlyn Guide). London: Hamlyn. (See also Maitland, 2000.)

Maitland, P. S., 1990. *Conservation of threatened freshwater fish in Europe*. Nature and Environment Series No. 46. Strasbourg: Council of Europe.

Maitland, P. S., 1992. The status of Arctic charr *Salvinus alpinus* (L.), in southern Scotland: a cause for concern. *Freshwater Forum* **2**: 212–227.

Maitland, P. S., 1995. *The ecological requirements of threatened and declining freshwater fish species in the United Kingdom*. Unpublished contract report to the Joint Nature Conservation Committee. Peterborough: JNCC.

Maitland, P. S., 2000. *Freshwater fish of Britain and Europe* (Hamlyn Guide). London: Hamlyn. (Revised and updated edition of Maitland, 1977.)

Maitland, P. S., 2003. The status of *Osmerus eperlanus* in England. *Research Report* No. 516. Peterborough, English Nature.

Maitland, P. S., 2004. *Keys to the freshwater fishes of Great Britain & Ireland, with notes on their distribution and ecology*. Windermere: Freshwater Biological Association.

Maitland, P. S. & Campbell, R. N., 1992. *Freshwater fishes* (The New Naturalist Library No. 75). London: HarperCollins.

Maitland, P. S. & Campbell, R. N., 1994. Fish Watching – a peer beneath the surface. *Birds – the magazine of the RSPB* (Winter 1994), pp. 63–66.

Maitland, P. S. & Lyle, A. A., 1991. Conservation of freshwater fish in the British Isles: proposals for management. *Aquatic Conservation* **2**: 165–183.

Maitland, P. S. & Lyle, A. A., 1996. The smelt *Osmerus eperlanus* in Scotland. *Freshwater Forum* **6**: 57–68.

Mann, R. H. K., 1971. The populations, growth, and production of fish in four small streams in southern England. *Journal of Animal Ecology* **40**: 155–190.

Mann, R. H. K., 1973. Observations on the age, growth, reproduction and food of the roach *Rutilus rutilus* (L.) in two rivers in southern England. *Journal of Fish Biology* **5**: 707–736.

Mann, R. H. K., 1996a. Species action plan: spined loach *Cobitis taenia*. Report to English Nature in association with the Environment Agency. *In* Mainstone, C. P. (Ed.), *Species management in aquatic habitats – Compendium of project outputs: species action plans and management guidelines*. Environment Agency R&D Project Record W1/640/3/M (1998).

Mann, R. H. K., 1996b. Environmental requirements of European non-salmonid fish in rivers. *Hydrobiologia* **323**: 223–235.

Marconato, A., Bisazza, A. & Fabris, M., 1993. The cost of parental care and egg cannibalism in the river bullhead, *Cottus gobio* L. (Pisces, Cottidae). *Behavioral Ecology & Sociobiology* **32**: 229–237.

Marlborough, D., 1970. The status of the burbot, *Lota lota* (L.) (Gadidae) in Britain. *Journal of Fish Biology* **2**: 217–222.

Mason, C. F., 1981 (Edn 3, 1996). *Biology of freshwater pollution*. Harlow: Longman.

Maxwell, H. E., 1904. *British freshwater fishes* (Woburn Library of Natural History). London: Hutchinson.

Melly, G., 2001. *Hooked!* London: Robson Books.

Miller, P. J., 1986. *A handguide to the fishes of Britain and Europe*. London: Treasure Press.

Miller, P. J. & Loates, M. J., 1997. *Fish of Britain and Europe* (Collins Pocket Guide). London: HarperCollins.

Mills, C. A. & Mann R. H. K., 1983. The bullhead *Cottus gobio*, a versatile and successful fish. *Annual Report of the Freshwater Biological Association* **51**: 76–88.

Mills, D. [H.], 1971. *Salmon and trout: a resource, its ecology, conservation and management*. Edinburgh: Oliver & Boyd.

Mills, D. [H.], 1989. *Ecology and management of Atlantic salmon*. London: Chapman & Hall.

Morris, D., 1954. The reproductive behaviour of the river bullhead (*Cottus gobio* L.) with special reference to the fanning activity. *Behaviour* 7: 1–31.

Morris, D., 1989. *Thomas Hearne and his landscape*. London: Reaktion Books.

Moss, B., Madgwick, J. & Phillips, G., 1996. *A guide to the restoration of nutrient-enriched shallow lakes*. Norwich: Broads Authority.

Muus, B. J. & Dahlstrøm, P. (Ed. A. Wheeler), 1971. *Freshwater fish of Britain and Europe* (Collins Guide). London: Collins. (Several subsequent editions with slight variations in title and by other publishers.)

Muus, B. J. & Dahlstrøm, P., 1981. *Sea fishes of Britain and northwestern Europe* (Collins Guide). London: Collins.

Myers, G. S., 1925. Introduction of the European bitterling (*Rhodeus*) and the rudd (*Scardinius*) in New Jersey. *Copeia* **140**: 20–21.

Nelson, J. S., 1976. (Edn 3, 1994). *Fishes of the world*. New York: John Wiley.

Norman, J. R., 1931. *A history of fishes*. London: Ernest Benn. (See also Greenwood, 1963.)

Ogle, D. H., 1998. A synopsis of the biology and life history of ruffe. *Journal of Great Lakes Research* **24**(2): 170–185.

O'Hara, K. & Penczak, T., 1987. Production of the three-spined stickleback, *Gasterosteus aculeatus* L., in the River Weaver, England. *Freshwater Biology* **18**: 353–360.

O'Maoileidigh, N. & Bracken, J. J., 1989. Biology of the tench, *Tinca tinca* (L), in an Irish lake. *Aquaculture and Fisheries Management* **20**: 199–209.

Park, D. W., 1972. *Freshwater fishes – a guide to the identification of those found in British waters*. Welwyn Garden City: British Ichthyological Association.

Paxman, J., 1994. *Fish, fishing and the meaning of life*. London: Michael Joseph.

Pecl, K., 1995. *Fishes of lakes and rivers*. Wigston, Leicester: Magna.

Pedroli, J-C., Zaugg, B. & Kirchhofer, A., 1991. *Atlas de distribution des poissons et cyclostomes de Suisse*. Neuchâtel: Centre suisse de cartographie de la faune.

Pennell, H. C., 1863. *The angler-naturalist: a popular history of British fresh-water fish*. London: G. Routledge.

Perrow, M. R. & Jowitt, A. J. D., 1997. *The habitat and management requirements of spined loach* Cobitis taenia. English Nature Research Reports No. 244. Peterborough: English Nature.

Perrow, M. R. & Jowitt, A. J. D., 2000. On the trail of the Spined Loach: developing a conservation plan for a poorly known species. *British Wildlife* **11**: 390–397

Perrow, M., Punchard, N. & Jowitt A. [J. D.], 1997. *The habitat requirements of bullhead* (Cottus gobio) *and brown trout* (Salmo trutta) *in the headwaters of selected Norfolk rivers: implications for conservation and fisheries*. Unpublished report by ECON (Ecological Consultancy to the Environment Agency). Norwich: University of East Anglia.

Phillips, R. & Rix, M., 1985. *A guide to the freshwater fish of Britain, Ireland and Europe*. London: Pan Books.

Pinder, A. C., 2001. *Keys to larval and juvenile stages of coarse fishes from fresh waters in the British Isles*. (Scientific Publication No. 60). Windermere: Freshwater Biological Association.

Pinder, A. C. & Gozlan, R. E., 2003. Sunbleak and Topmouth Gudgeon – two new additions to Britain's freshwater fishes. *British Wildlife* **15**: 77–83.

Porta, G. della, 1589. *Magiae naturalis libri XX in quibus scientiarum naturalium divitiae et deliciae demonstrantur*. Naples. (English translation, 1658, as *Natural Magick*. London.) (See: http://members.tscnet.com/pages/omard1/jportat5.html)

Potter, I. C. & Beamish, F. W. H., 1975. Lethal temperatures in ammocoetes of four species of lampreys. *Acta Zoologica. Stockholm* **56**: 85–91.

Quinn, T., 1991. *Angling in art*. London: The Sportsman's Press.

Rackham, O., 1986. *The history of the countryside*. London: Dent. (Several subsequent editions.)

Ramage, J., 1825. Account of a stickleback that was found with a leech alive in its intestines, July 1818. *Edinburgh Journal of Science* 3: 74.

Regan, C. T., 1911. *The freshwater fishes of the British Isles*. London: Methuen.

Robotham, P. W. J., 1978a. Some factors influencing the microdistribution of a population of spined loach, *Cobitis taenia*. *Hydrobiologia* **61**: 161–167.

Robotham, P. W. J., 1978b. The dimensions of the gills of two species of loach, *Neomachelilus barbatulus* and *Cobitis taenia*. *Journal of Experimental Biology* **76**: 181–184.

Rosecchi, E., Crivelli, A & Catsadorakis, G., 1993. The establishment and impact of *Pseudorasbora parva*, an exotic fish species introduced into Lake Mikri Prespa (north-western Greece). *Aquatic Conservation: marine and freshwater ecosystems* **3**: 223–231.

Rosecchi, E., Thomas, F. & Crivelli, A., 2001. Can life-history traits predict the fate of introduced species? A case study on two cyprinid fish in southern France. *Freshwater Biology* **46**: 845–853.

Salmon and Freshwater Fisheries Review Group, 2000. *Salmon and freshwater fisheries review*. London: MAFF. (Available via the internet at http://www.defra.gov.uk)

Schindler, O., 1957. *Freshwater fishes*. London: Thames & Hudson.

Scott, W. B. & Crossman, E. J., 1973. *Freshwater fishes of Canada*. Bulletin 184. Ottawa: Fisheries Research Board of Canada.

Serjeantson, S., Wales, S. & Evans, J., 1994. Fish in later prehistoric Britain. *In* Heinrich, D. (Ed.), Archaeo-Ichthyological Studies (Proceedings of the Sixth meeting of the International Council for Archaeozoology Fish Remains Working Group). *Offa* **51**: 332–339.

Shearer, W. M., 1992. *The Atlantic salmon: natural history, exploitation and future management*. London: Fishing News Books.

Simmons, I., Dimbleby, G. W. & Grigson, C., 1981. The Mesolithic. *In* Simmons, I. G. & Tooley, M. J. (Eds), *The environment in British prehistory*, pp. 82–124. London: Duckworth.

Sinha, V. R. P. & Jones, J. W., 1975. *The European freshwater eel*. Liverpool: Liverpool University Press.

Skelton, P. H. 1993. *A complete guide to the freshwater fishes of southern Africa*. Halfway House: Southern Book Publishers.

Smith, C. L., 1990. Moronidae. *In* Quero, J. C., Hureau, J. C., Karrer, C., Post, A. & Saldanha, L. (Eds), *Check-list of the fishes of the eastern tropical Atlantic (CLOFETA)* **2**: 692–694. Lisbon: JNICT; Paris: SEI; and Paris: UNESCO.

Smith, P., 2001. Can fish determine the conservation value of shallow lakes in the UK? *British Wildlife* **13**: 10–15.

Solomon, D. J., 1994. *Sea trout investigation – Phase I. Final Report*. Research & Development Note 318. Bristol: National Rivers Authority.

Steane, J. M., 1971. The medieval fishponds of Northamptonshire. *Northamptonshire Past and Present* **4**: 299–310.

Terofal, F. (Ed. and transl. G. Vevers), 1979. *British and European fishes: freshwater and marine species* (Chatto Nature Guides). London: Chatto & Windus.

Tesch, F.-W., 1977. *The eel. Biology and management of anguillid eels*. London: Chapman & Hall.

Thomas, M., 1998. Temporal changes in the movements and abundance of Thames estuary fish populations. *In* Attrill, M. J. (Ed.), *A rehabilitated estuarine ecosystem: the environment and ecology of the Thames estuary*, pp. 115–139. London: Kluwer Academic Publishers.

Thomson, J. M., 1951. Growth and habits of the sea mullet *Mugil dobula* (Gunther) in W. Australia. *Australian Journal of Marine and Freshwater Research* **4**: 41–81.

Thomson, J. M., 1953. Status of the fishery of sea mullet (*Mugil cephalus* L.) in Eastern Australia. *Australian Journal of Marine and Freshwater Research* **5**: 70–131.

Thomson, J. M., 1954. The Mugilidae of Australia and adjacent seas. *Australian Journal of Marine and Freshwater Research* **6**: 328–347.

Torbett, H., 1961. *The angler's freshwater fishes*. London: Putnam.

Treasurer, J. W. & Mills, D. H., 1993. An annotated bibliography of research on coarse and salmonid fish (excluding salmon and trout) found in fresh water in Scotland. *Freshwater Forum* **3**: 202–236.

van Utrecht, V. L., 1959. Wounds and scars in the skin of the common porpoise, *Phocaena phocaena* (L.). *Mammalia* **23**: 100–102.

van Waade, A., van den Thillart, G. & Verhagen, M., 1993. Ethanol formation and pH-regulation in fish. *In* Hochachka, P. W., Lutz, P. L., Sick, T., Rosenthal, M. & van den Thillart, G. (Eds), *Surviving hypoxia: mechanisms of control and adaptation*. Boca Raton, Florida: CRC Press.

Varley, M. E., 1967. *British freshwater fishes: factors affecting their distribution*. London: Fishing News Books.

Vik, R., 1954. Investigations on the pseudo-phyllidean cestodes of fish, birds and mammals in the Ånøya water system in Trøndelag. Part I. *Cyathocephalus truncatus* and *Schistocephalus solidus*. *Nytt magasin for zoologi* **2**: 5–51.

Volckaert, F. A. M., Hänfling, B., Hellemans, B. & Carvalho, G. R., 2002. Timing of the population dynamics of bullhead *Cottus gobio* (Teleostei: Cottidae) during the Pleistocene. *Journal of Evolutionary Biology* **15**: 930–944.

Walton, I., 1653. *The compleat angler or the contemplative man's recreation ...* . London: Richard Marriot.

Walton, I. & Cotton, C. 1676. *The complete angler* [with Part II by Charles Cotton on *The art of fly fishing, being instructions how to angle for a trout or grayling in a clear stream*]. London.

Wells, A. L., 1941. (Completely revised, Bagenal, T. B., 1970). *The Observer's book of freshwater fishes.* London: Warne.

Whealan, K. F., 1983. Migratory patterns of bream *Abramis brama*, L. shoals in the River Suck system. *Irish Fisheries Investigation Series A* **23**: 11–15.

Wheeler, A. C., 1969. *The fishes of the British Isles and north west Europe.* London: Macmillan.

Wheeler, A. [C.], 1974. Changes in the freshwater fish fauna of Britain. *In* Hawksworth, D. L. (Ed.), *The changing flora and fauna of Britain*, pp. 157–178. (Systematics Association Special Volume No.6). London: Academic Press.

Wheeler, A. [C.], 1977. The origin and distribution of the freshwater fishes of the British Isles. *Journal of Biogeography* **4**: 1–24.

Wheeler, A. [C.], 1978a. *Key to the fishes of northern Europe.* London: Warne.

Wheeler, A. C., 1978b. *Spotter's guide to fishes* (Usborne Pocket Book). London: Usborne.

Wheeler, A. C., 1979. *The tidal Thames: the history of a river and its fishes.* London: Routledge & Kegan Paul.

Wheeler, A. [C.], 1983. *Freshwater fishes of Britain and Europe.* London: Rainbow Books.

Wheeler, A. [C.], 1992. A list of the Common and Scientific Names of Fishes of the British Isles. *Journal of Fish Biology*, **41**, Suppl. A, pp. 1–37.

Wheeler, A. [C.], 1998. Field key to the freshwater fishes and lampreys of the British Isles. *Field Studies* **9**: 355–394. (Also published as a separate AIDGAP guide.)

Wheeler, A. [C.], 2001. Fishes. *In* Hawksworth, D. L. (Ed.), *The changing wildlife of Great Britain and Ireland*, 410–421. (Systematics Association Special Volume No.62). London: Taylor & Francis.

Wheeler, A. [C.] & Maitland, P. S. 1973. The scarcer freshwater fishes of the British Isles. I. Introduced species. *Journal of Fish Biology* **5**: 49–68.

Wheeler, A. [C.] & Newman, C., 1992. *The pocket guide to freshwater fishes of Britain and Europe.* Limpsfield: Dragon's World.

Whilde, A., 1993. *Threatened mammals, birds, amphibians and fish in Ireland. Irish Red Data Book 2: Vertebrates.* Belfast: HMSO.

Wiepkema, P. R., 1961. An ethological analysis of the reproductive behaviour of the bitterling (*Rhodeus amarus* Bloch). *Archives Neerlandaises Zoologie* **14**: 103–199.

Williamson, J., 1740. *The British angler: a pocket companion for gentlemen fishers... an exact description of the several kinds of fish that are found in the rivers and on the sea coasts of Great Britain.* London: J. Hodges.

Wilson, C. A., 1973. *Food and drink in Britain.* London: Constable.

Willughby, F., 1686. *De historia piscium libri quatuor....* London. (Facsimile edn, 1978. (Ed. K. B. Sterling), including the decorative engraved title page: *Francisci Willoughby Icthyographia ad Amplissimum Virum D^num^ Samuelem Pepys ...* . Manchester, New Hampshire: Ayer Publishers.)

Winfield, I. J., Fletcher, J. M. & Cragg-Hine, D., 1994. *Status of rare fish, a literature review of freshwater fish in the UK.* National Rivers Authority Research & Development Report No. 18. London: HMSO.

Winfield, I. J., Dodge, D. P. & Rösche, R., 1998. Introductions of the ruffe, *Gymnocephalus cernuus*, to new areas of Europe and to North America: history, the present situation and management implications. *In* Cowx, I. G. (Ed.), *Stocking and introduction of fish*, pp. 191–200. Oxford: Fishing News Books and Blackwell Scientific Publications.

Wintle, M., 2001. When Leney brought black bass into the UK. *Classic Angling* **14**: 15–17.

Wootton, R. J., 1976. *The biology of the sticklebacks.* London: Academic Press.

Wootton, R. J., 1984. *A functional biology of stickle-backs.* London: Croom Helm.

Wright, R. M. & Giles, N., 1991. The population biology of tench, *Tinca tinca* (L) in two gravel pit lakes. *Journal of Fish Biology* **38**: 17–28.

Xie, S., Cui, Y. & Li, Z., 2001. Dietary-morphological relationships of fishes in Liangzi Lake, China. *Journal of Fish Biology* **58**: 1714–1729.

Yarrell, W., [1835]–36. *A history of British fishes*, 2 vols. London: Van Voorst.

Youngson, A. F. & Hay, D., 1996. *The lives of salmon. An illustrated account of the life-history of Atlantic salmon.* Shrewsbury: Swan Hill Press.

Index

Note 1. Main entries for fish species are in **bold** type.

Note 2. For names of Lakes, Lochs, Reservoirs, Fens and other wetland areas, and of Rivers and Canals, *see under* **waterbodies** (1) and (2). These are further divided into British and Foreign waters including Inland Seas. Open Seas and Oceans are indexed separately under **waterbodies** (3).

Note 3. For British Acts of Parliament, Statutory Instruments and European Commission Regulations and other Conservation Directives, *see under* **legislation – national and international obligations.**

Note 4. Authors of literature cited in the text are indexed. For the titles of their works *see* Bibliography and further reading (pp. 163–169). For artists, *see* **fishes: in art.**